AF763052

MULTISPECIES INFORMATION SCIENCE

Multispecies Information Science offers a comprehensive overview of this subfield, shifting the focus beyond humans to explore the rich and complex realms of information and knowledge shared with animals, plants, microbes, technologies, and landscapes.

Bridging theory and practice, this volume traces the evolution of multispecies movement in information science, from Suzanne Briet's provocative question, 'Is an antelope a document?' to inclusive approaches informed by posthumanism, queer studies, disability studies, and postcolonialism. Penned by information scientists from around the world, the chapters span topics from the foundations of information to its pedagogy, design, technology, and services. Their ideas will help readers to discover fresh and compassionate avenues of participation and engagement with other species in their own information practices and research. By embracing a multispecies approach, this book aims to transform readers' understanding of information and life, recognising that this perspective is fundamentally an ethical turn.

Multispecies Information Science integrates insights from areas such as documentation, information behaviour, indigenous knowledge, and information technology. It facilitates dialogue between information scientists who are interested in a multispecies perspective. This book is essential reading for researchers, practitioners, and students in Library and Information Science. It also speaks to those in anthropology, environmental humanities, animal–computer interaction, and human–animal studies.

Niloofar Solhjoo is a Lecturer in the School of Information and Communication Studies at Charles Sturt University, Australia. Her research focuses on information experience, human–animal and multi-species relationships.

Global Perspectives on Library and Information Science

Global Perspectives on Library and Information Science highlights the wide range of innovative projects and approaches on offer within the LIS field today. Authored and edited by scholars and practitioners, volumes explore the vital work being done by library and information scientists all around the world.

The series contains global and regionally specific volumes and will provide students, researchers, and practitioners in the LIS field with diverse, unique insights into contemporary issues.

To submit a proposal or discuss an idea for a book in the series, please contact the Routledge Editor for Library, Archival and Information Science, Heidi Lowther (heidi.lowther@tandf.co.uk).

The following list includes only the most recent titles to be published within the series. A list of the full catalogue of titles is available at: https://www.routledge.com/Global-Perspectives-on-Library-and-Information-Science/book-series/GPLIS

Social Informatics
Edited by Noriko Hara and Pnina Fichman

Risk-Informed Information Practice
Ways of knowing in an uncertain world
Alison Hicks

Multispecies Information Science
Edited by Niloofar Solhjoo

MULTISPECIES INFORMATION SCIENCE

Edited by Niloofar Solhjoo

Routledge
Taylor & Francis Group
LONDON AND NEW YORK

First published 2026
by Routledge
4 Park Square, Milton Park, Abingdon, Oxon OX14 4RN

and by Routledge
605 Third Avenue, New York, NY 10158

Routledge is an imprint of the Taylor & Francis Group, an informa business

© 2026 selection and editorial matter, Niloofar Solhjoo; individual chapters, the contributors

The right of Niloofar Solhjoo to be identified as the author[/s] of the editorial material, and of the authors for their individual chapters, has been asserted in accordance with sections 77 and 78 of the Copyright, Designs and Patents Act 1988.

All rights reserved. No part of this book may be reprinted or reproduced or utilised in any form or by any electronic, mechanical, or other means, now known or hereafter invented, including photocopying and recording, or in any information storage or retrieval system, without permission in writing from the publishers.

For Product Safety Concerns and Information please contact our EU representative GPSR@taylorandfrancis.com. Taylor & Francis Verlag GmbH, Kaufingerstraße 24, 80331 München, Germany.

Trademark notice: Product or corporate names may be trademarks or registered trademarks, and are used only for identification and explanation without intent to infringe.

British Library Cataloguing-in-Publication Data
A catalogue record for this book is available from the British Library

ISBN: 978-1-032-95156-0 (hbk)
ISBN: 978-1-032-95155-3 (pbk)
ISBN: 978-1-003-58342-4 (ebk)

DOI: 10.4324/9781003583424

Typeset in Times New Roman
by SPi Technologies India Pvt Ltd (Straive)

To my husband,
who loves life and all its creatures.

CONTENTS

FIGURES

CONTRIBUTORS

Marcia J. Bates is Professor Emerita at the University of California at Los Angeles (UCLA) Department of Information Studies in the School of Education and Information Studies. She has written extensively on the nature of information and the information disciplines, information access, information-seeking behaviour, and user-centred design of information systems. She has also taught and consulted in these areas. Dr Bates is a Fellow of the American Association for the Advancement of Science, a recipient of the Association for Information Science and Technology's Award of Merit, as well as a recipient of the Frederick G. Kilgour Award for Research in Library and Information Technology. She was Editor-in-Chief of the seven-volume *Encyclopedia of Library and Information Sciences*, 3rd ed. (2010). Her three-volume *Selected Works of Marcia J. Bates* (2016) is available through online bookshops.

Ronald E. Day is a Professor of Information and Library Science at Indiana University Bloomington's Luddy School of Informatics, Computing, and Engineering. His approach is grounded in critical information studies and integrates rhetorical, conceptual, and historical analysis. Day is the author of several notable works, including The Modern Invention of Information: Discourse, History, and Power (2001) and Indexing It All: The Subject in the Age of Documentation, Information, and Data (2014)—the latter won the ASIS&T Best Information Science Book Award in 2015. His most recent work, Documentarity: Evidence, Ontology, and Inscription (2019), offers a philosophical lens on documentary evidence and inscription as modes of presence and judgement.

Björn Ekström is a Lecturer at the Swedish School of Library and Information Science, University of Borås. He explores human and nonhuman engagement with information in scholarly conduct, in scholars' relations with non-scholars and/or in environmental settings. He focuses on the situated interaction of information and how this interaction is relationally tied to large-scale phenomena. In his studies, Ekström combines socio-material practice theory approaches with qualitative and digital methods.

Steve Fuller holds the Auguste Comte Chair in Social Epistemology at the University of Warwick, UK. His latest book is *To Judge and To Justify: Profiles in the Academic Vocation* (Springer 2024). His next book is Media and the Power of Knowledge (Bloomsbury 2025). Also, in 2025, Springer is publishing Reforming the Governance of Science, a 25th anniversary revisiting of The Governance of Science (Open University Press, 2000). Originally trained in Politics and Philosophy,

Tim Gorichanaz is an Associate Teaching Professor in the College of Computing & Informatics at Drexel University. His areas of expertise are in information behaviour and experience, digital ethics, and human-centred design. His book *Information Experience in Theory and Design* was published by Emerald in 2020, and his other work has been published in venues such as the *Journal of Documentation*, the *Journal of the Association for Information Science and Technology*, and the Association for Computing Machinery conferences CHI and CSCW. He received his Ph.D. from Drexel University, his M.A. from New York University, and his B.A. from Marquette University.

Jenna Hartel is an Associate Professor at the Faculty of Information, University of Toronto. She is a theorist, methodologist, historian, and educator of Information Science. Her lifelong motivating question has been: What is the nature of information in the pleasures of life? To that end, Jenna has championed information research into 'higher things in life,' especially information-rich domains of serious leisure and hobbies. An enthusiast of multimodality and creative expression, she has introduced visual, photographic, and arts-informed methods into Information Science. Jenna's latest adventure, the YouTube channel INFIDEOS, features 100+ original, outrageously playful, educational videos about information science.

Dr Isto Huvila is a Professor of Information Studies at the Department of ALM (Archival Studies, Library and Information Science and Museums and Cultural Heritage Studies) at Uppsala University in Sweden. The prehistory of his work on archaeological information work and management goes back to the end of the last millennium. Huvila's primary areas of research include information and knowledge management, information work, knowledge organisation, documentation, and social and participatory information practices.

Monika Krakowska is an Information Science Scholar and Associate Professor at the Jagiellonian University, Kraków, where she heads the Unit of Information Interaction Studies at the Institute of Information Studies. She collaborates with the Centre for Higher Education Research of Jagiellonian University. She specialises in the study of information behaviour, with particular emphasis on information experiencing, the social dimensions of information practices, and the dynamics of hybrid information environments, including social media, augmented and virtual reality, extended reality, and user experience. Her research also explores information behaviour in multispecies contexts.

Finn Lees is a PhD student in Sociology at the University of Warwick, UK. His project, 'The Politics of Hunt Sabotage,' is an ethnographic study of direct action anti-blood sports groups in Britain and uses their practices to shed light on wider issues surrounding the politics of human–animal relations, social division in the ongoing culture wars, and the role of visual media in forming political subjectivities.

Yazdan Mansourian is a Senior Lecturer in the School of Information and Communication Studies at Charles Sturt University in Australia. He received his PhD in Information Science from the University of Sheffield in 2006. Since then, his research interests have been shaped around the cognitive and affective aspects of information behaviour in different contexts. Since 2017, he has focused on Information Behaviour (IB) in Serious Leisure. In this ongoing research programme, he explores the role of joy and other positive emotions in engaging people with hobbies, amateurism and voluntary activities and how joyful experiences inspire them to seek, share and use information. He also investigates to what extent people's engagement in joyful information activities can contribute to their overall wellbeing.

Dr Bhuva Narayan is a bookseller, editor, publisher, library educator, academic, and information researcher. And a lifelong dog lover. She is an Associate Professor in the School of Communication, Faculty of Design and Society at the University of Technology Sydney (UTS). Bhuva was Course Director for the Information and Knowledge Management Programme at UTS from 2013 to 2018 and Director of the Graduate Research Programme at the Faculty of Arts and Social Sciences from 2019 to 2024 and is currently Director of the Graduate Research Programme at the new Faculty of Design and Society at UTS. Her research areas are information behaviours, information avoidance, information access, user experience, human–computer interaction, personal information management, documentation studies, scholarly communication and open access. She is associate editor of the *Journal of the Australian Library and Information Association* (JALIA).

Ola Pilerot is Professor at the Swedish School of Library and Information Science, University of Borås. His research interests span several areas. In recent years, a significant part of his research has focused on issues related to the information practices of librarians. He has also written extensively on information literacies, including the recently published co-edited volume Information Literacy Through Theory (Facet Publishing, 2024). Theoretically, Pilerot usually works with some form of practice theory. It should also be mentioned that he lives with—among others—the cats Stig and Stina.

Erik Radio is an Associate Professor at the University of Colorado Boulder, where he serves as the Head of the Archival and Metadata Processing. His research spans metadata modelling, document theory, metadata auditing, and critical and semiotic approaches in Library and Information Science.

Niloofar Solhjoo is a Lecturer in the School of Information and Communication Studies at Charles Sturt University. Her work explores the ways humans, animals, and technologies experience information together. She has presented at conferences and published widely on human–animal information relationships, multispecies methodologies, and animal health information services. Her doctoral research was the first empirical study in information studies to adopt the multispecies turn, exploring the information experience within families.

Dirk van der Linden is an Assistant Professor in Animal–Computer Interaction at the School of Computer Science at Northumbria University and co-founder of the Tech4Animals Lab. He holds a BASc in Bioinformatics, an MSc in Information Science, and a PhD in Information Systems. His research focuses on the role of (information) technology in human–animal relations and computational animal behaviour analysis.

Magdalena Zych is an Assistant Researcher at the Unit of Information Research Methodology, Faculty of Management and Social Communication, Jagiellonian University, and collaborates with the Centre for Higher Education Research. Her research focuses on user experience, data storytelling, and cognitive approaches to relevance assessment, with interests in artificial intelligence and virtual reality.

FOREWORD

Are animals documents, documented, or ignored? How do our intellectual histories impact how we believe they use information? How do they impact the information we collect and produce about them? Should their information be included in our theories and systems? What are the challenges, barriers and ethics of including them? Can there be an ethical multispecies information studies? These are just some of the fascinating questions Niloofar Solhjoo and her co-authors address in this collection. They do so by asking what counts as information, who gets to define it, and how animals—and our relationships with them—challenge the boundaries of theory of information in LIS. Their different, and sometimes contentious, answers not only challenge the epistemologies underpinning the theory of information in LIS, but they also extend the field of animal studies by demanding that we take into account how human systems of information impact other animals (the environment and plants).

There is food for thought for animal studies scholars from various disciplines in here as the authors reflect on how living beings are often 'flattened' or reduced to texts through human attempts to visibilise and codify them. Drawing on rich and varied case studies, contributors show how removing an animal from their environment and 'translating' them into text through our information systems can occlude animal agency, presence, and ultimately legitimate a lack of empathy for them as individuals.

Conversely, they demonstrate how trying to think about the rich and various multispecies places information comes from, and how we might record it outside of current anthropocentric boundaries, leads to different conversations with the more-than-human world. Quite rightly, authors in this collection point out that simply focusing on other animals as objects of study does not make for a multispecies information study. So, instead of advocating for an 'add species and stir' approach, the authors here are at pains to show how the anthropocentrism inherent to LIS needs to be challenged and deconstructed. They do this through theoretical discussion and also through including a rich variety of case studies. And in so doing, they describe the wonder and joy of recognising multispecies worlds—a wonder and joy we might all want to hold on to as the basis for a critique of the nature/culture binaries that uphold animal oppression and abuse in our modern worlds. In this respect, the works here are foundational and offer something to those of us who (try to) use our work to identify and challenge the systems in place that maintain a world where other animals are often misused, abused, and oppressed by asking us to think about the knowledge and information networks in which our work is produced, reproduced and circulated.

As the field of animal studies grows, so do the disciplinary conversations around other animals. And this collection is a nice addition to this. One that, importantly, keeps a focus on why we might want to theorise about and include animals in our information systems differently. On how our own practices as academics can contribute to or hinder this aim. And on why—and how—it is imperative in current times to keep working toward the liberation of other species from the stranglehold that humans have on their lives and deaths.

Nik Taylor, Te Whare Wānanga o Waitaha | University of Canterbury

PREFACE

In Library and Information Science (LIS), when we care about something, it becomes part of our scholarly and practical domain. We work with and support many communities, individuals, artefacts, systems, disciplines, professions, and aspects of life. This diverse engagement drives us to question, explore, and imagine LIS as a fundamentally transdisciplinary field (Bates, 1999). Ultimately, our sphere of engagement with others shapes our own perspectives, capacities, knowledge, and skills as information professionals.

Since the 18th century, animals have been recognised as sentient beings (Duncan, 2006), compelling us to see ourselves as similar to, rather than in contrast with, nonhuman animals. Every species perceives the world through its own body and relations with its surrounding environment. If I were a dog, my knowledge would reflect a distinctly canine perspective. But what is unique about being human, however, is that our biases extend beyond individual perception to shape the very scientific systems we have constructed to catalogue, explain, and manage the world (Dunn, 2021). Yet, human life and its universe of documents—the core domain of information science (Bates, 1999)—is not about exerting control over the world, but about learning how to respond to it adaptively and meaningfully (de Waal, 2010).

The world is deeply interconnected, which dissolves the boundary between the individual human self and its environment. Each of us is a 'species multiple' (van Dooren et al., 2016), as our bodies are inhabited by a vast array of symbiotic and parasitic forms of life (Dunn, 2021). We exist within multispecies assemblages. Even a small action, like a butterfly flapping its wings, can trigger significant changes across the interconnected web of the external environment. The links between human activities such as global warming, deforestation, species loss and pandemics are undeniably evident and alarming. Yet, there have always been sustaining possibilities within more-than-human relationships, encompassing living beings (including humans), non-living technologies, natural processes, cultural meanings, and social relations. Such relationships invite careful, immersive engagement with the multiple nonhuman entities, such as animals, plants, fungi, microbes, viruses, stones, tornadoes, robots, and even gods that co-constitute the living universe.

Whenever I presented and spoke about the multispecies turn in LIS at various universities and conferences, I was often met with questions from information scholars and academics: *Does this mean our field and practice are becoming focused on animals and their information behaviour? If so, how are we different from disciplines that study animals and their behaviours?*

My response is: *the goal is not to erase the human centre, or to work solely with nonhuman animals; rather, the multispecies perspective seeks to reframe human-centredness in LIS*. As Cary Wolfe (2010) cautions, framing animal studies as a trans-discipline is not intended to create a super-discipline. Instead, each discipline (including LIS) can recognise that animals are not external to its theories, methods, and practices. They are already part of the world they study; we simply need to acknowledge them as such. This perspective emphasises that being central brings responsibilities to the wider web of relations that sustain us and to the 'red thread of information' (Bates, 1999) that we, as information professionals, perpetually seek (not only in the social texture of people's lives but in the wider texture of the more-than-human world).

We need to make ethical choices, yet those choices unfold within a world that exceeds us, our understanding, and our control (Bonelli & Walford, 2021). As an information scholar, professional, or student, ask yourself: *Do you truly consider and care for the many others who share the world with you? And if you do, how do you turn that awareness into meaningful action and engagement in your professional and academic practices?*

Niloofar Solhjoo
Charles Sturt University

References

Bates, M. J. (1999). The invisible substrate of information science. *Journal of the American Society for Information Science*, *50*(12), 1043–1050. https://doi.org/10.1002/(SICI)1097-4571(1999)50:12<1043::AID-ASI1>3.0.CO;2-X

Bonelli C., & Walford A. (2021). *Environmental Alterities*. Mattering Press.

De Waal, F. (2010). *The Age of Empathy: Nature's Lessons for a Kinder Society*. Crown.

Duncan, I. J. H. (2006). The changing concept of animal sentience, *Applied Animal Behaviour Science*, *100*(1–2), 11–19. https://doi.org/10.1016/j.applanim.2006.04.011

Dunn, R. (2021). *A Natural History of the Future: What the Laws of Biology Tell us about the Destiny of the Human Species*. Basic Books.

Van Dooren, T., Kirksey, E., & Münster, U. (2016). Multispecies studies: Cultivating arts of attentiveness. *Environmental Humanities*, *8*(1), 1–23, https://doi.org/10.1215/22011919-3527695

Wolfe, C. (2010). *What Is Posthumanism?* Minneapolis, MN: University of Minnesota Press.

ACKNOWLEDGEMENT

I would like to begin by thanking all the chapter authors who contributed to this volume, those who joined from the very start, those who responded to my invitations and open calls, and those who became involved later through presentations and discussions at conferences. I wish to acknowledge Marcia Bates, Ronald E. Day, Steve Fuller, Tim Gorichanaz, Jenna Hartel, Isto Huvila, Monika Krakowska, Finn Lees, Yazdan Mansourian, Bhuva Narayan, Ola Pilerot, Erik Radio, Dirk van der Linden, Björn Ekström, and Magdalena Zych. They generously and compassionately invested intellectual energy into this project, dedicating their chapters to the volume, reviewing each other's work, and helping to develop a cohesive collection. Special thanks to Jenna Hartel and Christopher Lueg for encouraging me from the very beginning to start and take the lead in editing this volume on the multispecies movement. I also thank potential chapter authors, Christopher Lueg, Melissa Ocepek, Spencer Lilley, Siyao Cheng, and Babasola Olugasa, who joined the process with exciting ideas but were unable to meet the schedule for the book.

Special thanks to Nik Taylor for graciously writing the foreword to this edited book, lending her voice and expertise to frame the work within the broader scholarly conversation. I am also grateful to the team who generously volunteered their time and insights to peer review chapters of this collection: Valtteri Anttonen, Roxanne Bodsworth, Nora Fagerholm, Jennifer Ford, Charlotte Hankin, Travis Holland, Rachel Hogg, Spencer Lilley, Maja Krtalić, Melissa Ocepek, Reijo Savolainen, and Kara White. Their thoughtful feedback supported me in the editing process and significantly strengthened the quality and coherence of the chapters.

I thank many nonhuman companions for their love and for guiding me in many directions. In particular, I thank Peter, my family dog; Otto, who was my companion during my time as an international doctoral student in New Zealand; and Pippi, the dog living next door, who shared her companionship with me when I first moved to Australia and began my new academic career. I also thank the many cats, dogs, birds, and other companion species who lived with and inspired the chapter authors of this volume.

I am grateful to my husband, Mohammad Beigi, with whom I share a deep love for animals and a long, loving partnership. Thanks as well to my parents, Azarmidokht Jalilnia and Ali Solhjoo, whose ongoing engagement with the progress of this volume and encouragement in my research, despite the long distance between us, motivated me to complete the volume on schedule.

Beyond human and nonhuman individuals, many discussions and presentations on this topic at conferences and research seminars shaped my framework for editing and writing this book. In particular, I acknowledge the Annual Meeting of the Association for Information Science and Technology (2023), the International Conference on Conceptions of Library and Information Science (COLIS) (2022, 2025), the School of Information Management at Te Herenga Waka—Victoria University of Wellington (2024), the Wee Kim Wee School of Communication and Information (2024), the School of Information and Communication, and the Creative Practice Circle at Charles Sturt University (2025) for providing opportunities to present, discuss, and refine ideas that informed this volume.

Finally, my thanks go to Heidi Lowther, Editor at Routledge Books; Manes Roy, Editorial Assistant; and the wider Routledge team for their support in bringing this topic to publication.

Niloofar Solhjoo
Charles Sturt University

ABBREVIATIONS

AI	Artificial Intelligence
ACI	Animal–Computer Interaction
HCI	Human–Computer Interaction
LIS	Library and Information Science
STS	Science and Technology Studies

SECTION 1
Multispecies Fundamentals

Niloofar Solhjoo

Multispecies studies encompass a variety of disciplinary and interdisciplinary approaches that emerged in the late 2000s. A key landmark was the publication of the influential manifesto 'The Emergence of Multispecies Ethnography' (Kirksey & Helmreich, 2010). However, the origins of multispecies studies trace back to a series of intellectual breakthroughs that challenged the human/animal binary and the anthropocentric assumptions in anthropology, science and technology studies (STS), and other fields. Humans and other animals are considered 'companion species' (Haraway, 2003), mutually shaping each other's natural and cultural evolution. Informed by post-human perspectives, new materialism, queer studies, disability studies, critical race theory, and postcolonialism, the multispecies lens highlights the importance of many other life forms as fundamental to human life, work, and research.

The edited book *Multispecies Information Science* is the first of its kind, focusing on realities beyond human centrality in LIS. It is not merely about information, its use, or its design in relation to other species. Rather, it is fundamentally concerned with the profound ethical responsibilities that information professionals have as companion species, coexisting and co-constituting the world. The chapters of this book cover a wide range of topics, from the foundations of information and documents to information technology, pedagogy, and professional practice. Penned by information scientists whose lives and scholarship are enriched by encounters with many other companion species, they offer unique insights and contributions to LIS.

The impetus for this volume began with a panel at the 2023 Annual Meeting of the *Association for Information Science and Technology* (ASIS&T) in London. During the panel, Dr Niloofar Solhjoo, Professor Christopher Lueg,

DOI: 10.4324/9781003583424-1

Associate Professor Jenna Hartel, Professor Steve Fuller, and Dr Dirk van der Linden introduced the multispecies turn as a paradigm shift in LIS (Solhjoo et al., 2023). The panel and meeting with the publisher at the conference sparked the initial ideas for this book. The chapter contributors were carefully selected and invited based on their expertise, prior publications, and personal interest in multispecies research. Chapter topics were determined through a balance of authors' expertise and interests, complemented by Dr. Solhjoo's editorial guidance and subject knowledge, to lay the groundwork across the historical, theoretical, methodological, ethical, educational, and practical dimensions of *Multispecies Information Science*.

This volume is intended for academics, students, and information professionals who are curious and compassionate about how the multispecies perspective can enrich the study and practice of LIS. It is well-suited as a central or complementary text in courses on the foundations of LIS, contemporary issues, information behaviour, and information design across iSchools. Whether you are a researcher, practitioner, educator, or student in librarianship, archival studies, documentation, information studies, knowledge management, information systems, or human–computer interaction, this book invites you to reflect on the ethical and relational responsibilities of sharing the world with other beings and to bring that awareness into your scholarly and professional work.

The book is structured to guide the reader from foundational concepts to application, organised into two main sections: *Fundamentals* and *Studies*. In this first section, the authors establish the foundations for the multispecies turn in LIS. It begins with 'The Footprints of Animals in Library and Information Science' by Niloofar Solhjoo and Jenna Hartel, which traces the historical and evolving role of nonhuman organisms in shaping LIS concepts, theories, and practices. This is followed by Marcia Bates's chapter, 'Researching Information Behaviour among Multiple Species,' which constructs a metatheory for studying information behaviour in shared ecologies, drawing on ecological psychology and the concept of the umwelt. Tim Gorichanaz then bridges disciplines in 'Information Experience for the Expanding Circle,' connecting expansions in information experience research with developments in animal–computer interaction. Ronald E. Day pushes our conceptual boundaries further in 'The Animal's Importance for Neo-Documentation,' arguing that a focus on animals precipitates a Copernican revolution in our very understanding of documentation.

The focus then shifts to education with 'When Animals Teach: Lessons for Information Pedagogy' by Solhjoo and Hartel, which introduces a transformative approach that integrates the agency and relationships of animals into the LIS curriculum. The theoretical foundations of the section are rounded out by an exploratory dialogue on 'The Ethics of Multispecies Information Research' between Steve Fuller and Finn Lees, who debate the ethical implications of documenting other species and designing multispecies research. Finally, Niloofar Solhjoo provides practical guidance in 'Methodological Pathways for Multispecies Information Science,' introducing a suite of qualitative, quantitative, and creative methods for conducting research within more-than-human assemblages.

Section 2, *Studies*, brings these theories to life through a diverse array of case studies and conceptual research. Together, the insights and practices in these sections invite readers to broaden their vision of LIS—revealing new opportunities for engagement, participation, and research across species, disciplines, and spheres of life.

References

Haraway, D. (2003). *The companion species manifesto: Dogs, people, and significant otherness*. University of Chicago Press.

Kirksey, S. E., & Helmreich, S. (2010). The emergence of multispecies ethnography. *Cultural Anthropology*, 25(4), 545–576. https://doi.org/10.1111/j.1548-1360.2010.01069.x

Solhjoo, N., Fuller, S., Hartel, J., Lueg, C., & Linden, D. (2023). Multispecies information science. *Proceedings of the Association for Information Science and Technology*, 60(1), 841–844. https://doi.org/10.1002/pra2.874

1

THE FOOTPRINTS OF ANIMALS IN LIBRARY AND INFORMATION SCIENCE

Niloofar Solhjoo and Jenna Hartel

Introduction

From ancient scrolls to digital libraries, cats and many other animals have long influenced how information professionals and scholars think about information and its use. The presence of cats in libraries—from the famed feline of Alexandria's Royal Library in Egypt (Stevens, 2004) to the robotic cats in public libraries of New South Wales, Australia (*The Guardian*, 2025)—is a long-standing entanglement between animals and those who work in information disciplines and professions. The library cat holds more than symbolic value and its utility as a working animal. It is a living presence within the library, shaping media portrayals, public perception, social interaction, trust, and institutional identity of libraries.

This chapter traces the evolving role of nonhuman entities, asking: *Where are the animals in Information Science?* From foundational ideas within Library and Information Science (LIS), like Suzanne Briet's antelope as document and Robert Pagès' gorilla as auto-document, to the story of Ishi (Nord, 2020), natureculture documents (Kosciejew, 2020), and various information theories that incorporate biological, evolutionary, and ecological frameworks, this chapter explores the multispecies movement in LIS.

Animals in the History of Information Science

Throughout the history of LIS, the world beyond the human has played a profound role in shaping the ideas and contributions of many scholars. Some of the most transformative thinkers have drawn inspiration from animals. Those who paid attention to and cared for the natural world—by loving, living, and

DOI: 10.4324/9781003583424-2

learning with other species (Solhjoo, 2024a)—have expanded the horizons of LIS in extraordinary ways.

By acknowledging humans as part of a broader network of life and recognising our entanglements with nonhuman species, we begin to see information not as a human construct but as something co-produced across species and more-than-human contexts. This view brings us into dialogue with what Bates (1999) called the 'red thread of information'—a connective force that runs through all living beings. Within this thread, animals appear not merely as objects of observation but as subjects that challenge, inspire, and enrich our understanding of documentation, information, and knowledge organisation.

The Antelope as Document

One of the most iconic and historical examples in LIS is the antelope. Known as Madame Documentation, Suzanne Briet was one of the first women to be employed as a librarian at the Biblioteque Nationale in Paris and was a leading figure in the theory of documentation (Buckland, 1995). In the mid-20th century, Briet famously declared, 'An antelope is a document if it is indexed in a zoo' (1951/2006). This deceptively simple statement marked a paradigm shift. With the perspective introduced by Briet, a document is no longer limited to written records. It can include any living or non-living entity (Figure 1.1).

FIGURE 1.1 Suzanne Briet's captive antelope as a document. (Image generated using Freepik).

This perspective is echoed in the work of Bruno Latour (1996), who used the example of a bird removed from the wild and displayed in a natural history museum. The bird gains informational value through processes of comparison, classification, and representation. Even Mike Buckland (2017a) recounts visiting a museum and initially finding the collection of preserved birds odd—until he started to see them as informational resources, like books on a shelf. This insight allowed Buckland to propose a unifying view of libraries, archives, museums, and databases as collection-based information centres, all grounded in the contextual transformation of materials into documents.

For Briet and so many others, documentation was not about the inherent qualities of an object but about its placement within systems of classification, study, and interpretation. A wild antelope in its natural habitat was not a document, but once brought into a zoo, catalogued, and studied, it entered the realm of documentation. As Buckland (2017a) notes, an antelope removed from its habitat becomes not just a displaced being but a different kind of document—one shaped by systems of meaning and interpretation. Therefore, human interactions are central to its recognition and function as a document.

What is missed in this view—though so critical in their own time—is the recognition of the animal's agency: Its capacity to be information, or to create information, and become informed on its own terms and in relation to its own world, without the presence and intervention of humans.

The Gorilla as Auto-Document

Not all documentation and definitions of information require human framing. This idea is first illustrated in Robert Pagès's example of a gorilla (featured by Mike Buckland). In 1946, Robert Pagès became a student in the documentation programme founded by Suzanne Briet (Buckland, 2017b). Pagès used an example of a gorilla to define an *auto-document*: Living and non-living objects that are just themselves, like a gorilla (Figure 1.2). Any document describing a gorilla is a secondary or derived document. For example, graphic documents are always about something—they are descriptive and, therefore, considered secondary to the objects or ideas they describe (Buckland, 2018).

What is fascinating about Pagès' gorilla is his emphasis on the self-evident nature of certain entities, which can be understood as documents without requiring human intervention. Pagès made a distinction between unique objects and specimens that represent a type. A gorilla is a distinct individual with a unique personality and life story. When viewed as a particular entity, it represents itself and is classified as a primary document. However, when the perceiver's interest shifts to the general characteristics of a group or population, these particulars become specimens—tokens of a type, sharing attributes with other members of the same class. In this case, it does not matter which specific gorilla is used; any member of the group suffices as a representative (Buckland, 2017a, 2017b, 2018).

FIGURE 1.2 Robert Pagès's gorilla as an auto-document. (Image generated using Freepik).

This duality reflects the complex relationship between individuality and classification in documentation. Together, Briet's antelope and Pagès's gorilla reveal a rich and ongoing dialogue within LIS about what counts as information, who gets to define it and in which context practices.

These iconic animals in our history challenge the boundaries of ontology and epistemology in LIS. Slowly, they have become part of a broader multispecies movement—one that invites us to reconsider not only how we classify the world but also how we live and work within it alongside other beings. This reorientation aligns with Ronald Day's (2018, 2019) development of the concept of *powerful particulars*, inspired by Pagès' gorilla as the auto-document. He challenges the traditional view of documentation as something humans impose onto passive subjects. Instead, Day focuses on beings that serve as self-evidential signs—entities that express and document themselves through their very existence and interactions. Higher-order organisms like gorillas, trees, and even bacteria participate in their own documentation. In Chapter 4 of this volume, Day (2026) discusses the challenges posed by conceptualising information as representational content about entities. He invites the readers to stop thinking of documentation as something humans do to the world and start seeing the world as documenting itself.

Animals in the Theory of Information Science

There are various recognised paradigms that have shaped LIS (see Bates, 2005). For example, Jenna Hartel (2019a) identified seven major turns in the field and speculated about future directions, notably highlighting the *animal turn* and the *family turn*. At the time, her insights about the future were grounded in emerging, independent, and non-mainstream research that was beginning to surface within LIS. Today, Hartel's vision is becoming a reality, as these turns are finding a central place within the discipline. *Multispecies Information Science*, serves as evidence of those shifts, alongside another forthcoming book edited by Narayan and her team on *Documenting Family Information and Life Histories* (Routledge).

What was particularly notable in Hartel's (2019a) vision was her advocacy for assemblages that extend beyond humans, incorporating them into more-than-human collectives. Bates (2022) echoed this inclusive vision in her keynote at the International Conference on Conceptions of Library and Information Science (CoLIS), where she presented a layered model outlining the research focal points of different information metatheories. These domains range from the mind and body to ecology and society dimensions.

In this section, we outline information metatheories, theories, and models that incorporate nonhuman and more-than-human entities—demonstrating that the multispecies turn has long been part of our field's evolving narrative.

Foundations of LIS Beyond the Human

Returning to Briet's famous example of the antelope in the zoo as a document, her groundbreaking ideas in the neo-documentation movement were strongly influenced by Paul Otlet. Otlet, an earlier theorist, envisioned a universal knowledge network that included not just books but also physical objects, images, and living organisms (Otlet, 1934). Otlet believed organisation was a natural law shared across life forms—humans, animals, and machines.

In North America, Pierce Butler played a pivotal role in shaping early LIS. His seminal textbook, *An Introduction to Library Science* (1933), established foundational concepts for the field, particularly through his framework of scholarship, which addressed the entire documentary universe. This framework included information institutions, the words of books, the content of thoughts, and human experience. At its heart was *cosmic phenomena*, a broad category encompassing not only human endeavours but also biological processes and animal behaviour. Butler specifically emphasised the significance of social and gregarious animal species, integrating them into his vision of the natural and intellectual order. While his treatment of animals was not exhaustive, their inclusion at the core of his conceptual model highlights their importance in his understanding of knowledge and information systems.

Thinkers such as Briet, along with others discussed in this chapter, employ a rhetorical strategy that begins with biological and natural specimens to unsettle and expand established conceptions of information phenomena (Day, 2001). For instance, Bonnie Nardi and Vicki O'Day's *Information Ecologies*

(1999) introduced ecological thinking into LIS by using a biological metaphor to explore the integration of information technology into society. They defined an information ecology as 'a system of people, practices, values, and technologies in a particular local environment' (Nardi & O'Day, 1999, p. 49). These ecologies are seen as dynamic ecosystems where people, tools, and values co-evolve. Within such environments—like libraries—they identified *keystone species* such as librarians, who play a critical role in maintaining the balance and functionality of the entire system.

Naturally, a strong connection exists in people's minds between information and the biosphere (Hartel & Savolainen, 2016). For instance, the iSquare study (Hartel et al., 2018) aimed to expand and diversify existing conceptions of information through arts-informed, visual methods using the draw-and-write technique (Hartel, 2019b). More than 5,000 research participants were asked, 'What is information?'—and invited to respond with a drawing. Unsurprisingly, many responses featured conventional artefacts (e.g., a book, a computer) or abstract patterns and symbols (Hartel, 2014). However, organic metaphors—such as the Earth, web, tree, cloud, seedling, sun, and mountains—were among the ten most prevalent visual representations, as shown in Figure 1.3. These natural forms appear especially effective in capturing elusive information structures: Global and vast (Earth), hierarchical and branching (tree), networked and egalitarian (web), or remote and shifting (cloud). Beyond the philosophical and analytical literature that defines information within information science (Bates, 2010), the foremost source domain of information is, in fact, nature itself (Hartel & Savolainen, 2016). The iSquare study concluded that such ecological visual metaphors are more universally accessible than many verbal definitions; they lend a meaningful sense of depth, holism, and ancient continuity, and they affirm information science's role in the urgent conversation about our planet's future.

Advancements in artificial intelligence have reshaped how humans relate to non-living entities—entities that increasingly exhibit lifelike behaviour and participate actively in decision-making. AI not only influences how we analyse information but also transforms what is known and how it is known, even in multispecies contexts, suggesting mutual insights across domains. On the frontier of artificial intelligence, Bender et al.'s (2021) celebrated critique casts large

FIGURE 1.3 Nature was a major source domain of pictorial metaphors for information, namely: Earth, web, tree, cloud, seedling, sun, and mountain. (Redrawn from Hartel & Savolainen, 2016).

language models as *stochastic parrots* which are capable of sounding like humans but are devoid of real meaning. Here, we will not reproduce their comprehensive argument. Of greater interest is the choice of a flamboyant animal metaphor; the appearance of a bird emoji in the article title (🦜); Bender's penchant for wearing a parrot necklace; and a related article centred on a thought experiment involving an octopus (Bender & Koller, 2020).

In response to AI's epistemic challenges, Calzati (2025) critiques the traditional DIKW (Data–Information–Knowledge–Wisdom) pyramid for its linear, anthropocentric logic. Rather than treating data as neutral input processed upwards through discrete layers, he argues that all forms of knowing are situated and relational. Calzati reconceptualises knowledge as emerging within ecosystems—where humans, animals, machines, and material objects co-create meaning. In his model, agents (human or AI) are not static processors but open-ended and situated within specific contexts. Calzati's ecosystemic framework reveals how data, information, and knowledge are deeply interwoven through relational and environmental entanglements.

All the foundational frameworks and viewpoints discussed here challenge the human-centred view of information science, showing instead that animals, humans, and even material objects co-participate in the creation of information and knowledge. Reflecting this shift, in the next section, we turn to one of the key areas of information science, which is increasingly highlighting the multispecies dynamics at the core of how information is experienced and acted upon.

The Animal Turn in Human Information Behaviour

Human information behaviour is not simply a modern phenomenon. In *Information Behavior: An Evolutionary Instinct*, Spink (2010) describes information behaviour as an evolving instinctive ability in early human species. She provides an overview of the human species (e.g., early hominins, Homo sapiens, and modern humans), tracing the development of their information artefacts and information behaviour. Spink's focus on different human species and their biological and cultural evolution is distinctive. For instance, *information foraging* (Pirolli & Card, 1998, which parallels how animals forage for food. This includes the proximal perception of *information scent*, which helps individuals decide whether a particular source or path is worth pursuing (Pirolli & Card, 1998).

Marcia Bates has played a significant role in broadening this evolutionary perspective in LIS. She provides a nuanced vocabulary for understanding information as it manifests across all living beings. She suggests that information, like biological traits, evolves as an adaptive internal resource across all organisms, humans, animals, and plants (see also Bates, 2005; Madden, 2004).

Drawing on Susantha Goonatilake's (1991) three levels of information (i.e., genetic, neural-cultural, and exosomatic), Bates (2006) expands the concept of information to include natural and represented information, encoded and embodied information, and various experiential forms of information. These include enacted, expressed, embedded, recorded, and trace information shared among all species. Marcia Bates' work lays the groundwork for further exploration of nonhuman information behaviour. She expands on her work in Chapter 2, developing her metatheoretical framework through the multispecies lens and outlining approaches for researching interspecies information behaviour (Bates, 2026).

Other scholars have similarly developed evolutionary frameworks for understanding information behaviour. A notable example is Christopher Lueg's (2014) model, which conceptually illustrates how human information behaviour is grounded in perception, emphasising that only perceived information can influence behaviour. He argues that information behaviour is deeply intertwined with our cognitive and embodied capacities, which have co-evolved with our environments. In this view, what we do not perceive does not influence our information behaviour. Lueg also notes a human tendency to prioritise visual information over auditory or other sensory stimuli, suggesting that if something cannot be sensed, we may struggle to register it as information.

In a more recent contribution, Lueg (2024) extends his ideas through the multispecies lens. He outlines three key benefits of the animal-inclusive approach in information studies and design. First, recognising the diversity of perceptual and bodily configurations across species could enrich our understanding of human information behaviour. Second, it would enable information professionals to contribute more meaningfully to biodiversity issues through interdisciplinary collaboration. Third, it would provoke deeper theoretical inquiry into the very nature of information, questioning whether information must be interpreted by humans or if it can be meaningful by virtue of biological existence alone. To support this shift, Lueg proposes practical solutions such as creating digital libraries of animal perspectives, or *viewsonas*, which simulate how animals perceive human-altered environments.

Contemporary research continues to embrace animals as agents in information studies. Weber (2020) highlights the agency of library cats, while Grenersen (2016) discusses knowledge transfer between Sámi people and their reindeer. In the study with researchers working with environmental sound recordings experience, Stewart-Robertson (2025) shows how knowledge production emerges through entangled human and nonhuman actors. For instance, the meaning of a sound depends on research goals and environment: For a bat researcher, isolating a single species' call makes other sounds noise, whereas for an ecologist studying biodiversity, the entire wide cluster of noise is the signal, representing a healthy, complex soundscape. Thus, context determines which elements of a recording are valued as information (pp. 576–577), blurring distinctions between human researchers, recording devices, and animals as co-actors in knowledge production.

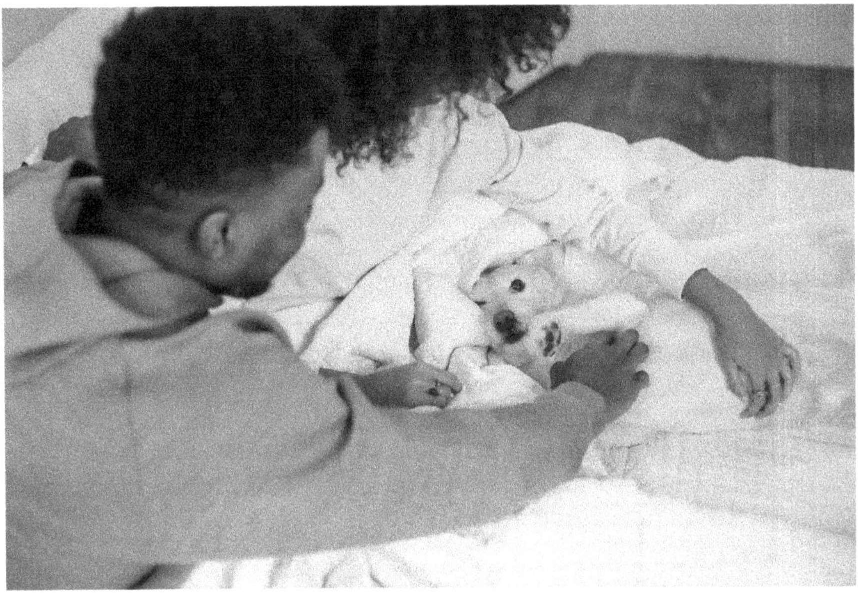

FIGURE 1.4 The human–environment–animal assemblage nurtures relational ways of knowing through love and empathy. (Photo by Andres Ayrton, on Pexels).

Moving a step forward, Solhjoo investigated how information is distributed and enacted across humans, cats, dogs, and material objects as *multispecies families* (Solhjoo et al., 2024). Her findings suggest that love is a form of relational knowing across companion species. Dogs and cats share affective, non-verbal information with their humans. This occurs in shared socio-physical environments, such as through touch, body posture, facial expressions, or internal chemosignals, while resting together on the bed (Figure 1.4). These interactions shape and inform human-animal co-becoming within the family (Solhjoo, 2024b). Apart from human information experiences, the study also revealed that animals draw on their past experiences (e.g., trauma), routines (e.g., dinner time, walks), and instincts (e.g., hunting) to inform their training and learning with humans (Solhjoo, 2024c).

Beyond this, Solhjoo emphasised how humans and animals co-create shared communication systems through playful activities. Much of this communication is non-verbal. For instance, *touch talk* (petting, holding, brushing, etc.) exemplifies receiving, acquiring, and interpreting sensory and cognitive-affective information. Such embodied exchanges occur in what she calls *coming together spaces* (e.g., couch, bed, home office, dog parks), where sensory material generates shared meanings in intimate contexts (Solhjoo, 2024b). In these interactions, humans and animals often mirror one another's emotions, fostering mutual empathy and deeper relational bonds. This requires abandoning an anthropocentric standpoint and embracing knowing as an ongoing process that is both environmental and agential, grounded in love and empathy rather than a tendency to control the environment (Solhjoo, 2024b).

It is not possible to become fully informed without others (Solhjoo et al., 2024). Humans learn from animals, animals learn from humans, and both are shaped by their environment. As Lueg (2014) argues, information is deeply rooted in perceptual and embodied capacities. Loving, living, and learning with animals (Solhjoo, 2024a) enhances what humans (and even cats and dogs) can perceive as information, mutually influencing each other's capacities to sense, respond to, and interpret the world. According to Solhjoo et al. (2024), the human–environment–animal assemblages challenge the view of information as solely a human phenomenon by highlighting the vital roles that many nonhumans play in co-creating everyday information behaviour. Information flows *in-between* beings (through bodies and shared environment), capturing the relational and participatory sense-making that unfolds when companion animals and their human companions co-create meaning (p. 722).

Rethinking Information Science: Nonhumans vs. Humans

When the boundaries between human/animal, nature/culture, and living/non-living begin to blur, we must also question long-held divisions between users and sources. When we speak of information sources in LIS, do we refer only to things, systems, or even animals stripped of subjectivity? And when we speak of individuals, do we mean only humans as information users or providers?

The Human as Document

The concept of the human has historically been constructed in opposition to animality, nature, and embodied existence (Wolfe, 2010). However, the line between human and nonhuman has not always been drawn the same way. The split between nature and culture, between the civilised and the wild, is ideological. For instance, throughout history, especially during colonisation, that line was often twisted to serve power. Colonisers denied the humanity of Indigenous peoples, treating both people and land as objects to be owned or controlled (Radio, 2020). Unlike the colonial framework, which often positioned Indigenous peoples as closer to nature to justify their dehumanisation, Indigenous knowledge and lifeways strive to listen to what the land and animals are already communicating, recognising their voices and intelligences as integral to their knowledge systems.

A vivid example of this is found in the Ute people's understanding of bears as ancestral relatives. By observing the emergence of hibernating bears from their caves in March, the Ute were able to break winter camps and search for

FIGURE 1.5 The Bear Dance, illustrating the entanglement of human culture with nature. (Photo by Rare Book Division, The New York Public Library, 1845, on Unsplash).

food. They believe their primal ancestors were bears and that, upon death, ancient Utes transformed into bears, establishing an enduring kinship between the species. Each spring, they perform a ten-day ceremonial dance known as the Bear Dance, timed to coincide with this natural event (Figure 1.5). The dance also facilitates human courtship and marriage, paralleling the way female bears search for mates in the spring (Reed, 1896). Through these practices, humans learn the rhythms of life by observing, listening, and participating in the natural world and the culture of animals, rather than seeking to dominate or separate from it. For further discussion on the multispecies perspective in Indigenous knowledge systems, see Chapter 5 (Solhjoo & Hartel, 2026).

The challenge to construct binaries between human/animal and nature/culture becomes even more urgent when considering how documentation practices have historically reinforced these boundaries. Take the story of Ishi, the last known member of the Yahi people (Nord, 2020). Ishi became a human document through the convergence of genocide, anthropological inquiry, and early recording technologies. Nord draws on Frohmann's (2012) concept of *documentality*—which emphasises function, historical contingency, complexity, and agency—to show how Ishi's representation fluctuated: From museum exhibit to storyteller to an individual with personal, undocumented moments. Ishi's case reveals a core ethical dilemma in documentation: The risk of reducing a living being to a fixed representation, severed from the complex web of relationships and environments that shape it. This flattening can obscure suffering and diminish the potential for empathy and understanding (Nord, 2020).

In the face of these challenges, information science must move beyond merely cataloguing and recording nature to actively engaging with the silences and absences that continue to shape multispecies knowledge. In Chapter 14 of this volume, Erik Radio (2026) explores how the Anthropocene has brought about the extinction of countless species and, with them, the erasure of multispecies knowledge. In this chapter, Erik Radio argues that documents should not merely memorialise loss but engage with absences, or forms of life, and meaning that persist beyond recognition even when they're not directly visible or easily defined.

The Seed as a Natureculture Document

Seeds carry the imprint of how nature shapes culture, holding within them the memory of multispecies pasts and the possibility of futures yet to unfold. Marc Kosciejew (2020) introduces the concept of the *natureculture document* using the example of seeds preserved in seed banks (Figure 1.6). He draws inspiration from Donna Haraway's (2003) idea of natureculture, which describes the inseparable relationship between biological life (nature) and human meaning-making

FIGURE 1.6 Marc Kosciejew's seeds as natureculture documents. (Image generated using Freepik).

systems (culture). Seeds carry genetic memories that are vital for ecological resilience and food security. Seeds connect the past (e.g., the history of natural evolution and human agricultural practices) with the future (e.g., new scientific exploration). They are living documents in seed banks that embody crucial knowledge for both science and survival (Kosciejew, 2020). In this sense, seeds act as active agents in the documentation process .

It is even possible to base an entire information theory on fungi. Neither plant nor animal, but occupying their own kingdom, fungi are fascinating and featured in seminal multispecies studies (Sheldrake, 2020; Tsing, 2015). Concerned with magic mushrooms, James Kent's (2010) *Psychedelic Information Theory: Shamanism in the Age of Reason* has few parallels in the information science literature and is exceptionally comprehensive. It begins by explaining how mushrooms that contain the hallucinogen, psilocybin, alter cognition to create novel *psychedelic information* in the mind. It then describes the social worlds and information flows between shamans and their followers, who communicate through embodied modalities of chanting, drumming, and dancing. Ultimately, Kent surveys the socio-cultural impacts of psychedelic information upon politics, music, and art worldwide. The concluding chapter in the book outlines *information genesis* since the beginning of time and relates the irregular leaps involved in psychedelic information to quantum computing and artificial intelligence. Kent spoke at the Canadian Association for Information Science conference on the panel *Psychedelic Information Theory: A Systematic Exploration of Information During Altered Consciousness* (Hartel et al., 2024), and Hartel (2023) provides a multimedia introduction to his vision.

Conclusion

In this chapter, we reflect on several examples that invite us to rethink LIS through a multispecies lens, from Briet's antelope to Pagès' gorilla as auto-document, from Butler's integration of biological processes into the documentary universe to Lueg's ideas of species-specific sensory modalities in digital libraries. If we embrace the idea that animals are not just objects to be used and studied (or simply ignored), but they are knowers, users, providers, and contributors in decision-making, then the very definitions of document, information, and knowledge are changed.

We encourage reflection on a truly multispecies turn. LIS should not be limited to organising the world in conventional, human-centric formats. Instead, they should be about being in conversation with the more-than-human world, listening to what animals and many others are saying. Information lives in the bark of a tree, in the migration of birds, and in the whisper of seeds, questioning many human ways of defining and using information. To succeed in multispecies information science, we need to reflect on how our concepts, theories, methods, and practices account for the construction of the human–animal divide, without becoming solely about animals (Wolfe, 2010). LIS already operates across diverse domains, applying an informational lens (Bates, 1999) that connects the arts and sciences, work and leisure, and the civilised and the wild. To advance, LIS must move beyond its deliberate human-centred constructions while preserving its disciplinary strengths and engaging critically with the human–animal divide.

We hope the stories in this chapter inspire you to question how animals are included or excluded in your own perspective and professional work. This reflection applies across areas such as classification, information retrieval, information experience, information literacy, information policy, information design, information architecture, knowledge organisation, and artificial intelligence. At the same time, we caution careful engagement that sustains multispecies information science as a reflexive perspective. We do not advocate for a multispecies turn that dominates LIS, but rather as one approach among many for reimagining the discipline.

References

Bates, M. J. (1999). The invisible substrate of information science. *Journal of the American Society for Information Science*, *50*(12), 1043–1050. https://doi.org/10.1002/(SICI)1097-4571(1999)50:12<1043::AID-ASI1>3.0.CO;2-X

Bates, M. J. (2005). Information and knowledge: An evolutionary framework for information science. *Information Research*, *10*(4), n4. http://InformationR.net/ir/10-4/paper239.html

Bates, M. J. (2006). Fundamental forms of information. *Journal of the American Society for Information Science and Technology*, *57*(8), 1033–1045. https://doi.org/10.1002/asi.20369

Bates, M. J. (2010). Information. In M. J. Bates & M. N. Maack (Eds.), *Encyclopedia of Library and Information Sciences* (3rd ed., Vol. 3, pp. 2347–2360). CRC Press.

Bates, M. J. (2022). A proto-paradigm for information science research. In *Proceedings of CoLIS, the 11th. International Conference on Conceptions of Library and Information Science*, Oslo, Norway, May 29 – June 1, 2022. Information Research, 2 7. https://doi.org/10.47989/colis2201

Bates, M. J. (2026). Researching information behaviour among multiple species. In N. Solhjoo (Ed.), *Multispecies Information Science*. (pp. 22–38) Routledge.

Bender, E. M., Gebru, T., McMillan-Major, A., & Shmitchell, S. (2021). On the dangers of stochastic parrots: Can language models be too big? In *Proceedings of the 2021 ACM Conference on Fairness, Accountability, and Transparency*, 610–623. https://doi.org/10.1145/3442188.3445922

Bender, E. M., & Koller, A. (2020). Climbing towards NLU: On meaning, form, and understanding in the age of data. In *Proceedings of the 58th Annual Meeting of the Association for Computational Linguistics*, 5185–5198. https://doi.org/10.1177/183693911303800208

Briet, S. (1951/2006). What is documentation? In R. E. Day, L. Martinet, & H. G. B. Anghelescu (Eds. & Trans.), *What Is Documentation? English Translation of the Classic French Text* (pp. 9–46). Lanham, MD: Scarecrow Press. (Original work published 1951)

Buckland, M. (1995). The centenary of Madame-documentation, Briet, Suzanne 1894–1989. *Journal of the American Society for Information Science*, *46*(3), 235–237.

Buckland, M. (2017a). Reflections on Suzanne Briet. In *Fondements épistémologiques et théoriques de la science de l'information-documentation: Actes du 11e colloque ISKO*. France.

Buckland, M. (2017b). Before the antelope: Robert Pagès on documents. *Proceedings from the Document Academy*, *4*(2), 1–17, Article 6. https://doi.org/10.35492/docam/4/2/6

Buckland, M. (2018). The expansion of bibliography: Dog-stone, antelope, and evidence. (Lecture). https://escholarship.org/uc/item/0bj042km (Archived at https://web.archive.org/web/20250402054804/https://escholarship.org/uc/item/0bj042km)

Butler, L. P. (1933). *An Introduction to library science*. University of Chicago Press.

Calzati, S. (2025). An ecosystemic view on information, data, and knowledge: Insights on agential AI and relational ethics. *AI & Ethics*. https://doi.org/10.1007/s43681-025-00665-0

Day, R. (2026). The animal's importance for neo-documentation: A 'copernican revolution' in understanding information. In N. Solhjoo (Ed.), *Multispecies Information Science*. (pp. 54–61) Routledge.

Day, R. E. (2001). *The Modern Invention of Information: Discourse, History, and Power*. Southern Illinois University Press.

Day, R. E. (2018). Auto-documentality as rights and powers. *Proceedings from the Document Academy*, *5*(2), 3.

Day, R. E. (2019). *Documentarity: Evidence, Ontology, and Inscription*. MIT Press.

Frohmann, B. (2012). The documentality of Mme Briet's antelope. In J. Packer & S. B. Crofts Wiley (Eds.), *Communication Matters: Materialist Approaches to Media, Mobility and Networks* (pp. 173–182). Routledge.

Goonatilake, S. (1991). *The Evolution of Information: Lineages in Gene, Culture and Artefact*. Pinter.

Grenersen, G. (2016). Documentation, information, and the animal connection. *Proceedings from the Document Academy*, *3*(2), 15. https://doi.org/10.35492/docam/3/2/15

Haraway, D. J. (2003). *The Companion Species Manifesto: Dogs, People, and Significant Otherness*. University of Chicago Press.

Hartel, J. (2014). An arts-informed study of information using the draw-and-write technique. *Journal of the Association for Information Science and Technology*, *65*(7), 1349–1367. https://doi.org/10.1002/asi.23121

Hartel, J. (2019a). *Turn, turn, turn. In Proceedings of CoLIS, the Tenth International Conference on Conceptions of Library and Information Science*, Ljubljana, Slovenia, June 16–19, 2019. *Information Research, 24*(4), paper colis1901. http://InformationR.net/ir/24-4/colis/colis1901.html

Hartel, J. (2019b). Draw-and-write techniques. In P. Atkinson, S. Delamont, A. Cernat, J. W. Sakshaug, & R. A. Williams (Eds.), *SAGE Research Methods Foundations*. SAGE Publications. https://doi.org/10.4135/9781526421036

Hartel, J. [@INFIDEOS]. (2023, December 11). Psychedelic information theory (Episode #6 of An Archipelago of Information Science) [Video]. *YouTube*. https://www.youtube.com/watch?v=fgo3GoR07rE https://www.youtube.com/watch?v=0

Hartel, J., Kent, J. L., Gorichanaz, T., & Munro, K. (2024, June 4–7). Psychedelic information theory: A systematic exploration of information during altered consciousness [Conference session]. *The 52nd Annual Conference of the Canadian Association for Information Science / l'Association canadienne des sciences de l'information (CAIS/ACSI)*, Online. https://journals.library.ualberta.ca/ojs.cais-acsi.ca/index.php/cais-asci/article/view/1840

Hartel, J., Noone, R., Oh, C., Power, S., Danzanov, S., & Kelly, B. (2018). The iSquare protocol: Combining research, art, and pedagogy through the draw-and-write technique. *Qualitative Research, 18*(4), 433–450. https://doi.org/10.1177/1468794117722193

Hartel, J., & Savolainen, R. (2016). Pictorial metaphors for information. *Journal of Documentation*, *72*(5), 794–812. https://doi.org/10.1108/JD-07-2015-0080

Kent, J. L. (2010). *Psychedelic Information Theory: Shamanism in the Age of Reason*. PIT Press/Supermassive, LLC.

Kosciejew, M. (2020). The concept of natureculture document: A conceptual exploration of seeds, embodied information, and unconventional records. *Journal of Contemporary Archival Studies*, *7*(1), 15.

Latour, B. (1996). Ces réseaux que la raison ignore: Laboratoires, bibliothèques, collections. In M. Baratin & C. Jacob (Eds.), *Le pouvoir des bibliothèques: La mémoire des livres en Occident* (pp. 23–46). Albin Michel.

Lueg, C. (2014). Characteristics of human perception and their relevance when studying information behavior. *Journal of Documentation*, *70*(4), 562–574.

Lueg, C. (2024). Why a future-oriented information science discipline should embrace an 'animal turn'. In *Proceedings of the 52nd Annual Conference of the Canadian Association for Information Science / l'Association canadienne des sciences de l'information (CAIS/ACSI)*, 4–7 June 2024. https://cais2024.ca/talk/21.lueg/21.Lueg.pdf

Madden, A. D. (2004). Evolution and information. *Journal of Documentation*, *60*(1), 9–23. https://www.emerald.com/insight/content/doi/10.1108/00220410.10516626/full/html

Narayan, B., Zijlema, A., Reyes V., & Kennan M. A. (Eds.) (forthcoming). *Documenting Family Information and Life Histories: Transnational Perspectives*. Routledge.

Nardi, B. A., & O'Day, V. L. (1999). *Information Ecologies: Using Technology with Heart*. MIT Press.

Nord, M. I. (2020). Ishi, Briet's antelope, and the documentality of human documents. *Proceedings from the Document Academy*, *7*(1), 8. https://doi.org/10.35492/docam/7/1/8

Otlet, P. (1934). *Traite' de documentation*. Editiones Mundaneum.

Pirolli, P., & Card, S. K. (1998, May). Information foraging models of browsers for very large document spaces. In *Proceedings of the working conference on Advanced visual interfaces* (pp. 83–93), New York, United States. https://doi.org/10.1145/948496.948509

Radio, E. (2020). Documents for the non-human. *Journal of Critical Library and Information Studies*, *3*(1), 1–18. https://doi.org/10.24242/jclis.v3i1.108

Radio, E. (2026). Massive unseen companions and the documentary project. In N. Solhjoo (Ed.), *Multispecies Information Science*. (pp. 222–235) Routledge.

Rare Book Division, The New York Public Library. (1845). *The Bear Dance*. https://qa-digitalcollections.nypl.org/items/3739ee40-c6bc-012f-3256-3c075448cc4b

Reed, V. Z. (1896). The Ute bear dance. *American Anthropologist*, *9*(7), 237–244.

Sheldrake, M. (2020). *Entangled Life: How Fungi Make Our Worlds, Change Our Minds & Shape Our Futures*. Random House.

Solhjoo, N. (2024a). Information experience within multispecies families: Understanding shared life with companion cats and dogs (doctoral dissertation). Open Access Te Herenga Waka–Victoria University of Wellington. https://openaccess.wgtn.ac.nz/articles/journal_contribution/The_In-between_Information_Experience_within_Human-Companion_Animal_Living/24986298

Solhjoo, N. (2024b). The red thread of information across more-than-human love. *Journal of Librarianship and Information Science*. https://doi.org/10.1177/09610006241255418

Solhjoo, N. (2024c). Knowing within multispecies families: An information experience study. *Journal of Information Science*. https://doi.org/10.1177/01655515241268845

Solhjoo, N., & Hartel, J. (2026). When animals teach: Lessons for information pedagogy. In N. Solhjoo (Ed.), *Multispecies Information Science*. (pp. 61–76) Routledge.

Solhjoo, N., Krtalić, M., & Goulding, A. (2024). The in-between: Information experience within human-companion animal living environments. *Journal of Documentation*, *80*(4), 701–729. https://doi.org/10.1108/JD-08-2023-0160
Spink, A. (2010). *Information Behavior: An Evolutionary Instinct*. Springer.
Stevens, Norman D. (2004). Myeo: The First Library Cat. *Published Works*, 48. http://digitalcommons.uconn.edu/libr_pubs/48
Stewart-Robertson, O. (2025). 'It's a wide cluster of noise': experiencing and describing information from environmental sounds. *Information Research an International Electronic Journal*, *30*(CoLIS), 574–582. https://doi.org/10.47989/ir30CoLIS51895
The Guardian. (2025, May 28). *Therapeutic Robot cats 'a Stroke of Genius' at Blue Mountains library, Australia*. https://www.theguardian.com/technology/2025/may/28/therapeutic-robot-cats-blue-mountains-library-australia
Tsing, A. L. (2015). *The Mushroom at the End of the World: On the Possibility of Life in Capitalist Ruins*. Princeton University Press.
Weber, A. S. (2020). Die Bibliothekskatze – Wie sie ihren Platz in der Bibliothek von der Mäusefängerin zur Botschafterin für Storytelling behauptet [The library cat – How it asserts its place in the library from mouser to storytelling ambassador] (Master's thesis). University of Vienna. https://phaidra.univie.ac.at/detail/o:1393225
Wolfe, C. (2010). *What Is Posthumanism?* University of Minnesota Press.

2

RESEARCHING INFORMATION BEHAVIOUR AMONG MULTIPLE SPECIES

Marcia J. Bates

Introduction

In this chapter, information will be discussed in relation to the life activities of humans, other animals, and plants. A model will be presented of the relationships among these living things that can form an effective basis for empirical research on multispecies information behaviour in information studies. I will argue that using the framework of ecological psychology constitutes a powerful way to integrate our understanding of these relationships for research purposes.

In what follows, the key definitions and concepts will be presented and integrated to form the argument. In sequence, a suitable definition of information will be described, then the ecological psychology paradigm developed by J.J. Gibson and, to a lesser extent, Roger Barker will be elaborated, followed by a discussion of the key concepts of 'affordance' and 'behaviour setting.' When these concepts are integrated in a single model, they will form a good basis for research on multispecies information behaviour.

Information

A book that addresses information in relation to humans as well as other living species, both animal and plant, must necessarily define information differently and more broadly than in the conventional discussions of information. Information cannot just be defined as facts, news, or intelligence, or as records, documents, or books, as we might commonly assume. When a male bird dances in the jungle and flashes his plumage in the presence of a female of his species, he is signalling and communicating something meaningful to him and her—about

DOI: 10.4324/9781003583424-3

his age (mature enough to mate but not over the hill), about his health (abundant and vivid plumage signalling his robustness), and other information that will play a role in whether she responds to the dance or not. Information is surely involved here: Several males may dance, and she will select one or more of them, *based on (her experience of) their dances*. Communication has occurred; something was conveyed and acted upon.

How do we define information in this context? Indeed, how do we define information in the many contexts that we recognise as somehow involving information but which do not fit into the narrower definitions referenced above?

In fact, one could argue, as I have, that a field with the concept of information at its centre, as ours is, *should* define information as broadly as necessary to encompass all the environments and circumstances in which the word is used for the purposes of our research. The definition provided here for the term is as broad as it needs to be, but not too broad. It does not sweep up senses of non-information within its orbit.

Let us begin with the absolute basics, then build up our understanding of information from there. Matter is any physical substance, that is, something that has volume and is capable of being weighed. Energy is the capacity to do work, that is, to cause movement or change.

In the process of existing through time, matter and energy fall into various conformations, patterns, and structures. *The pattern of organization of matter and energy is information*. (Bates, 2005b, 2006).

When we speak, sound waves shape the air and cause the eardrums of listeners to vibrate. These vibrations are converted into a series of processes until they reach the auditory centre of the brain, where further nervous system processes give meaning or recognition to the sound.

When rocks fall in a rockslide on an airless planet, the rocks break and roll downhill in various shapes and conformations. On one part of the planet, the ground may be flat; on another part, there may be piles and heaps of rocks sticking above the surface of the planet. The information of the flat part of the planet differs from the information of the rockslide part because the shapes of the matter differ in the way just described. If some living species comes along that can detect, that is, see or feel or in some other way perceive the flat part and the rocky part, then that species has represented the information of the planet's surface within its nervous system.

To say this, however, is not to say that information is *identical* with the physical materials or waves that make up the pattern of organisation. *The information is the pattern of organization of the material, not the material itself*. As Wiener has said, 'Information is information, not matter or energy' (1961, p. 132).

Mars Rovers can film and transmit visual information representing the surface of the planet back to living beings on Earth, so the means of detection need not be in a living creature, but only living creatures make use of said information.

In all these examples, matter and energy are involved. *The pattern of organization* of that matter and energy is information. It is not necessary for a sensing organism to perceive information for the information to exist. The configuration of rocks on the example planet is there, regardless of any sensing being. It just is. But when that sensing organism comes along, using its perceptual and information-processing apparatus, it can detect distinguishable aspects of that environment and use that detected information to act in life.

Life and Information

When sensing beings came along, they found ways to identify differences and distinctions in their environments that were useful to them—the presence of things they could consume as food, for example. They developed means of representing their informational perceptions in their nervous systems and then used that information for their purposes. All animals use information to get around safely in the world. For example, because we can see, we do not bump into a chair when we walk into a room. It is important to understand, however, that this pattern of using information to protect and support successful living holds true all up and down the line, from the smallest to the largest animals. Even plants have various means of sensing their environment and responding to threats (Chamovitz, 2012; Sheldrake, 2020; Wohlleben, 2017; Yong, 2022).

If we imagine people having a picnic on a blanket on the grass in a park, we can think of all kinds of different ways the perceiving beings in the park are experiencing the picnic. A dog sniffs all the different kinds of food and is eagerly waiting to be tossed a piece of something good. As his sense of smell is so much more powerful than that of the humans sitting around the blanket, he also notices the faint whiff of deodorant coming from the human being sitting next to him and the smelly feet of the barefoot boy who is his master. One of the humans radiates a subtle sick odour, detectable by the dog, that indicates some underlying illness not yet perceived by the humans.

Among the human beings, one woman is embarrassed because the food she brought is not as fancy as that brought by the other people at the picnic. Her friend, sitting across from her, notices her friend's chagrin and catches her eye to smile reassuringly. None of the other humans at the picnic pick up on this exchange between the two women, though the one woman's dog does sense the emotional discomfort of his mistress.

Off to the side, some scout ants are marching rapidly towards the group. Strong and varied smells from the picnic augur a bonanza for the ants. They locate the direction of the food and start scent-marking the trail with 'trail pheromones,' so that others from the nest can follow them and reap the harvest.

Overhead, a bird watches warily as these big human animals are getting too close to his growing chicks in their nest in the tree. If someone tries to climb the tree and get closer to the nest, the bird will dive-bomb them to try to scare them away. The bird has four-colour vision, in contrast to humans' three-colour vision, and can see colours that humans have no experience of and cannot imagine.

All the animals at the picnic are seeing, hearing, smelling, and feeling different things. The oldest woman of the group of humans feels the hard earth beneath the blanket acutely, while most of the others present do not notice it. Each animal present at the picnic is attending to different things in their environment (or in their mind). The scene is a symphony—or cacophony—of sensory input and processing by all the animals present.

Every animal has what biologist Jakob von Uexküll (1934/2010) called an 'umwelt,' that is, a surrounding perceptual life-world that is distinctive to that animal's experience. This umwelt is associated with whatever senses the animal has and its size and relationship to the surrounding world. The world as experienced by each species (and sometimes, especially among humans, by each individual within a species) is dramatically different from one to another. The umwelt of the ant—with the blades of grass looming high overhead—is very different from the umwelt of the humans or of the dogs or of the birds. We often unthinkingly assume that the environment experienced by other animals is just like the environment we experience, that we live in a sort of stage setting that is the same for us all. However, we cannot understand those other animals fully until we remind ourselves how different their umwelt is for them. The cat has poorer colour vision than we have, but better nighttime vision and a more acute sense of smell. Our interactions with a cat will be different, therefore, than we might otherwise assume.

We need to think about the umwelt experienced by the animals we are studying and compare that to our own umwelt. Human beings have retinal cone photoreceptors for red, green, and blue and hence are called trichromatic. Most domestic animals are dichromatic. They see blue but have trouble distinguishing red, green, and yellow. 'Color vision in these species is analogous to that of "color-blind" humans, who are usually missing their red and green cone population' (Ofri, 2018).

So, colour vision is better in humans than in cats and dogs, yet 'study results show that cats can detect light that is 6 times dimmer than the lowest detectable threshold of humans' (Ofri, 2018).

Further,

> Pets are able to watch TV, however it is less perfect than it is for humans. In pets the response rate of the cones in the retina is slower than in people. Pets have a difficult time fusing the TV images into a continuous image. For pets the images are flickering rather than continuous.
>
> *(West Suburban Humane Society, accessed 11/23/2024)*

> Cats have an excellent sense of smell, but dogs are the champions of the sense of smell. Dogs can smell as far as 20 kilometers (12.4 miles) away. Their hearing is also more sensitive than that of humans.
>
> *(University of Adelaide, accessed 11/23/2024)*

With our excellent colour vision, easily movable skull and body, and our ability to stand a couple of metres above the ground's surface and see a long way away, humans have evolved to emphasise vision more than smell and hearing. Much more could be said about differences in umwelts among the various species being studied, but the main point has been made that we must remain aware of these marked differences when studying interactions among humans and other animals.

The noted ethologist, Konrad Lorenz, wrote:

> Life is an eminently active enterprise aimed at acquiring both a fund of energy and a stock of knowledge, the possession of one being instrumental to the acquisition of the other.
>
> *(Lorenz, 1977, as cited in Kovac, 2007, p. 70)*

Information gathering and processing will happen differently with each species and individual, but, by Lorenz's principle, all living species need energy to live, and they need information to find energy. They use both in order to do all the other things, such as evade predators and reproduce, that animals do to survive.

Ladislav Kovac, in his 'Fundamental principles of cognitive biology' (2000), has worded it somewhat differently:

> There is a universal characteristic of any living system to sense relevant features of its surroundings and to react appropriately upon them in order to preserve its own permanence, its *onticity*.
>
> *(p. 54)*

> Hence, life from its very beginning is a cognitive system: the self-copying molecule, pursuing its onticity in the world…is already a *subject* facing the world as an object.
>
> *(p. 59)*

Plants, not usually being mobile, have more limited opportunities to find things in their environments that meet their needs, but they nonetheless are able to find sunlight and nutrients through growth and movement of roots and above-ground structures. Some plants have underground root networks that communicate information and resources (Wohlleben, 2017).

Depending on its evolutionary history and resulting life needs, every living thing goes about attempting to meet its needs—exercising its onticity—usually with large amounts of information involved in the process. Animals described above have different sensory modes, different umwelts, different types of nervous system storage, and different types of communication, but they are all loci of intense information processing and use.

Pervasiveness of Information in Life

Some people have reacted to this definition/description of information by saying, 'Oh, but then information is everything! Such a definition is of no use—too broad, too general.' But information is *not* everything. Rather, it is the *pattern of organization of everything*, of matter and energy. Information is not matter and energy itself, but rather the pattern of organisation of it. Let us return to the example rocky planet above. The rocks have fallen, rolled downhill, broken, and ended up in a particular configuration. This rock on the left, as we view it, rises up to a metre above the ground, while that rock on the right is two metres high.

Suppose two astronauts drive their exploration vehicle past the point where the two rocks sit. Later, one astronaut can say to the other,

> Do you remember that place where there were two rocks, one twice as high as the other? Well, I think that would be the best place to return to and start searching for that dried-up lakebed we saw on the photo of this planet.

The rocks did not intend to form themselves into a particular configuration. They did not *care* about being one metre or two metres high. They just happened to end up this way. Human beings, with sophisticated visual and memory capabilities; however, can *identify the resulting configuration of the rocks*, label it, remember it, and talk about it to each other. Later, at a debriefing session, the astronauts can say, 'We found this location where there is a one-meter rock and a two-meter rock—here are the coordinates we recorded. We recommend that point for starting the search.'

Think of all the uses of information in the above paragraph. Suppose both astronauts remember seeing the two rocks—they have the memory of them in mind—and have language, an extremely sophisticated form of information transmission, to talk about it. The first astronaut noted a feature of the rocks—one being twice the height of the other—that she could use to jog the other's memory. They record the coordinates of the location with sophisticated technology that will enable them to return to the exact location they want. They have in mind and on some recording device the appearance of the dried-up lakebed, which they have made as their objective for the exploratory sortie. Vision, memory, communication, and recording devices are all means of capturing and using information—a pervasive part of all animals' lives.

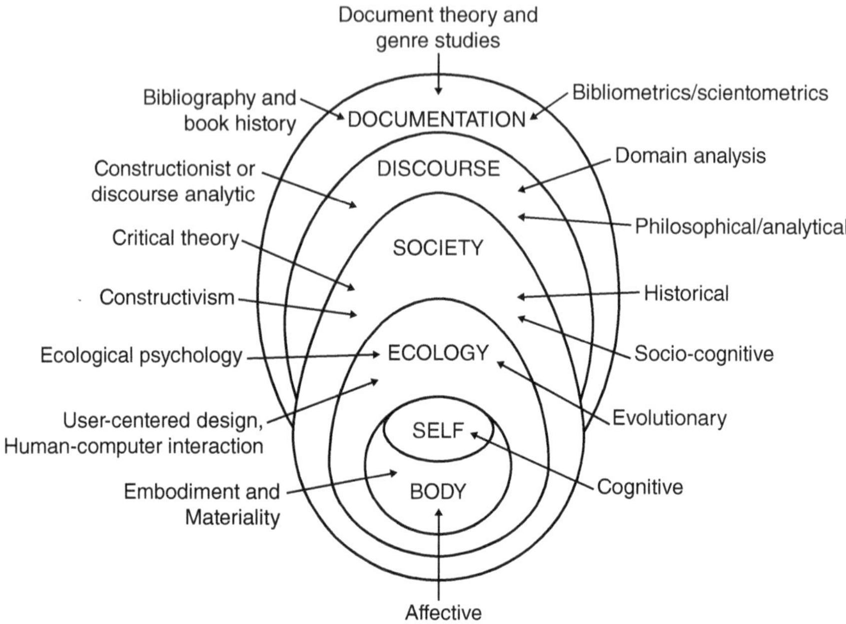

FIGURE 2.1 Research focal points of metatheories in information studies (Bates, 2022). Some metatheories draw from multiple domains.

Models for Multispecies Research on Information

This broad definition of information, which arguably covers all situations where we as researchers would want to talk about information, enables us then to consider how we in information studies should model our discussions and research about information. At a recent conference, I presented a model of the broad research domains involving information in life (Bates, 2022). These domains are the self, the body, ecology, society, discourse, and documentation. Then, having identified them, I suggested what metatheories in information studies most closely mapped to the layers of the model. See Figure 2.1.

These positions will not be reviewed here; readers familiar with the debates in information science around these approaches will recognise many of them. (See also Bates, 2005a.) Here, however, I wish to propose one of these approaches as being particularly suited to the study of information seeking and use across multiple species. That metatheory is ecological psychology.

Ecological Psychology

Ecological psychology is a branch of psychology in which the interaction and integration of people with their environment is closely studied. Most of conventional psychological research focuses on the individual human and the operation of mind and behaviour, largely independently of what is going on around the person. Standard psychological research seeks to identify invariants in mind and behaviour that are native to human beings, whereas ecological psychology starts from the premise that human behaviour cannot be fully understood apart from ecological conditions and influences.

We human beings are intensely absorbed in our social connections and relationships, in the many things that emanate from our social character, such as language, culture, politics, society, and technology, as well as interpersonal relationships. It can be easy to think of the rest of our environment as simply background noise, hardly noticed unless some anomalous event or experience draws our attention to it.

However, our species started out like other mammalian species, in a physical environment with all the life possibilities and threats that other animals are subject to: Finding food, mating, escaping predators, etc. There were no houses, nation-states, pretty clothes, cookbooks, or automobiles. There was solely the natural world and our conspecifics, and we had to make do within that environment. Over millions of years, our forebears adapted to that environment. During the thousands of years of our history as modern humans, our species, *Homo sapiens sapiens*, lived as hunter-gatherers.

The animal companions we live with today are rather like those we had during most of our history. They cannot make cookbooks or automobiles. The equipment they bring to our relationships with them is largely the same original equipment we had before our cognitive capabilities exploded and produced the remarkable efflorescence of social and technological developments we have enjoyed in recent millennia. To say this is not to preclude the very real possibility that we and our animals have evolved, or even co-evolved, over the time we have been associated with them. My point here is that in studying the human relationship with animals and plants, it may be most revealing to start with the baseline natural equipment and environments that we share as fellow animals.

Real-World Vision

Ecological psychology is rooted in the research on perception carried out by J.J. Gibson in the mid-20th century. He was most interested in vision. He gradually came to believe that the standard models of perception developed in the rest of the psychological field were wrong in many ways. He describes the standard psychological model of visual perception as follows: From the object in the world, an image is formed in the retina. That retinal image is sent to the brain. There, various operations are carried out on that sensory image. The result is full consciousness of the object and its meaning (Gibson, 1979/2015), p. 241). This model assumes that the incoming information is ambiguous and must be resolved into its meaning by processing in the brain.

In this standard model, the human being and the object in the world are separate and are linked by the ability to perceive the object through the eye. Gibson starts from a very different orientation. For him, a human is an animal on the Earth. At any given moment, the animal feels itself in a spot on the Earth. Movement leads to an awareness of now moving over the Earth, to a new spot elsewhere. A sense of being in a place is utterly fundamental to Gibson's orientation. Vision is a part of that sense-of-place-ness.

Traditional research on vision included having a person hold their head and eyes in a fixed position, with the aid of props that held the head in place. Setting up experiments in this way enabled researchers to focus exactly and only on what was happening in the eye when a person was looking at something. However, Gibson pointed out that this fixed, rigid seeing was utterly unlike real-world seeing. For the human being, real vision takes place in the context of five nested systems, each one contributing to the overall experience of perceiving visually. The first system is the eye itself, which we usually think of as the sum total of the visual apparatus. However, for vision to work, the eyes have a second system, muscles that enable the eyes to move back and forth in unison. Eyes are very mobile *within the head*, thus allowing for greater precision in viewing than would be possible otherwise. Third, the two eyes set slightly apart enable *binocular vision*, that is, 3D vision, which gives far more information than monocular sight would to the human being. Fourth, 'the eyes in a mobile head that can turn comprise an organ for the pickup of ambient information' (1979, p. 234). Fifth, 'the eyes in a head on a body constitute a superordinate organ for information pickup over paths of locomotion.' (1979, p. 234). Sensing kinesthetically our movement and direction is a fundamental part of our visual experience; it gives us a sense of where we are in the world. Thus, for Gibson, human vision is what happens when we move over the Earth and use the eyes, the eye muscles, binocular sight, movement of the head, and movement of the body in a coordinated supersystem to take in information.

Gibson contrasts the conventional research on vision with his own perspective. 'Color, form, location, space, time, and motion—these are the chapter headings that have been handed down through the centuries, but they are not what is perceived' (1979, p. 229).

Information Pickup and Affordances

As noted earlier, Gibson rejected the standard model of vision in the field of psychology, in which the person receives a retinal image of an object and sends it to brain processing centres, where the ambiguity of the original input is resolved and clarified into having meaning for the recipient. To Gibson, the original environment possesses enough specific detail to enable the viewer to directly recognise the object without a lot of mental processing. A person gets new information by just picking up on the changes in their pre-existing awareness. He calls this 'information pickup' (1979, p. 244). He notes that information pickup is 'the concurrent registering of both persistence and change in the flow of structured stimulation' (1979, p. 228). 'To perceive is to be aware of the surfaces of the environment and of oneself in it' (1979, p. 244). In sum, in Gibson's conceptualisation of perception, perceiving is a dynamic exploratory process. The human being is continually aware of where it is in relation to the environment and continuously 'coperceives' self and place in that environment.

> Perceiving gets wider and finer and longer and richer and fuller as the observer explores the environment. The full awareness of surfaces includes their layout, their substances, their events, and their affordances. Note how this definition includes within perception a part of memory, expectation, knowledge, and meaning—some part but not all of those mental processes in each case.
>
> *(1979, p. 244)*

Our study of humans and animals interacting and exchanging information is enriched by seeing all the animals, including us, within a common ecological context, each perceiving and understanding in our distinct species-specific ways.

Note how Gibson's description fits with our observation in information science of browsing behaviour. (See also discussion in Bates, 2007). The paradigmatic example of browsing is standing at a magazine stand, glancing around at the covers of the magazines. One is simultaneously orienting to the environment of the stand (are the magazines grouped by subject, by size, by educated vs. tabloid, or randomly?) and the individual covers (what topic catches my eye? Which shall I pick up first?). As Gibson notes, perception gets wider and richer and fuller as one orients to the stand, glancing over different parts of it, picking up on catchy story titles, etc. This instance is the quintessential one of an animal moving over the Earth, orienting itself in a particular environmental context.

So, what do people see with Gibson's real-world vision instead? People are continuously 'coperceiving' both the world and themselves; they are constantly aware of both themselves and the world they are in, the place where they are situated. According to Gibson, rather than primarily seeing colour, texture, etc., they are seeing *affordances*. 'The *affordances* of the environment are what it *offers* the animal, what it *provides* or *furnishes*, either for good or ill' (1979, p. 119). When in the forest, the most rudimentary affordance would be that the ground will hold me up if I take a step. Or, is there a root or a hole that will trip me up and cause me to fall? In the forest, we are continuously picking up information that informs us about our status as living animals. Are those berries on that bush edible? Is that cubbyhole big enough to hide from the rain? As noted above, information pickup involves 'the concurrent registering of both persistence and change in the flow of structured stimulation' (Gibson, 1979, p. 228). 'Perceiving is not the passive reception of sensory input, but it is a function of a dynamic, exploratory system' (Heft, 2001, p. 177).

In our current civilised form of living, we are also still constantly aware of the affordances, negative or positive, offered to us by our surroundings and social situation. Will the interviewer hire me for the job? Will the store have the model I am looking for? Am I driving too fast on a rainy street? Can the doctor help with my health problem? Will the love interest text me back? In sum, will these situations *afford* me what I want/need? We are constantly assessing our status, avoiding pitfalls and taking up opportunities afforded by our social and physical circumstances.

Behaviour Settings

So far, the discussion has been about the individual seeing and assessing his/her environment and responding to affordances. Another ecological psychologist, Roger Barker, worked at a level above that of the individual. He studied the settings within which human beings act. He analysed the relationships between people and the environments they used and shaped. As creative, complex, social animals, we often select physical environments, then modify them a little or a lot to harmonise with the way we intend to use them. Biologists call this niche construction (Odling-Smee, Laland, & Feldman, 2003; see also Bates, 2018a, pp. 260–262). To play baseball, we select a large flat space, measure and mark it with bases, and provide dugouts for teams and stands for watching the games—all before a single human being has arrived, actually, to play baseball. To teach children, we build classrooms with a large number of characteristics that make the teaching and learning that goes on in them more effective and satisfying. An observer familiar with American culture who has never seen a classroom in a school in Kansas may nonetheless speedily recognise the room as a classroom and be able to describe many activities that likely would take place in it.

Barker described these places as 'behaviour settings,' that is, 'dynamic structures comprised of standing patterns of behaviour and milieu' (Heft, 2001, p. 281).

According to Barker & Wright (1955),

> The structure of the behaviour pattern and the structure of the milieu in a behaviour setting are seen to be congruent, to fit, to be synomorphic. The criterion of this synomorphism is perceptual; it is directly seen (Barker & Wright, 1955, p. 46, quoted in Heft, 2001, p. 288). … The meaning of a behaviour setting, that is, what kind of a setting it is and thus what kinds of activities are appropriate in that place, resides in perceived synomorphic relations between milieu features and behaviour.
>
> *(Barker & Wright, p. 53, quoted in Heft, 2001, p. 288)*

This view of behaviour settings can be seen to harmonise nicely with Gibson's affordances. The human being, like other animals, is continually reading its environment to identify and take advantage of affordances. Humans, however, being very social animals and having created many social activities that they share with other humans, also create behaviour settings, milieux, which they have adapted to be particularly suitable to the conduct of certain activities that the humans intend to carry out and share. Once these behaviour settings have been created, humans can make use of the affordances designed into those settings (see also Heft, 2001, p. 289).

Note the phrase 'directly seen' in the quote from Barker & Wright. People often recognise the nature of a behaviour setting first from its direct, physical appearance and layout. When walking into a building or location, one immediately recognises the various features that make the setting a good match for the behaviours intended for that setting. One knows where one is on the Earth because, directly and initially before all else, one sees the features of that setting that mark it as a schoolroom, baseball field, church sanctuary, etc. This recognition is so automatic that social scientists are apt to ignore it and concentrate only on the social meaning imposed on the milieu. But we are often cued first and foremost by the physical behaviour setting itself. I think we can assume that animals are even more directly and intensely cued by their physical environment, because they have fewer and less developed emotional, cognitive, or other layers of response influencing their experience of a new milieu.

Barker's insight about the importance of behaviour settings arose in the context of studying the daily lives of children in a Midwestern town. In following multiple children through the day, he and his colleagues noticed that the children's behaviour was much more congruent with and predictable from the locale they were in than with any other factor in their situations (Barker, 1968, p. 152).

Ecological Psychology in the Study of Multispecies Information Behaviour

How can the theory of ecological psychology be integrated with the theory and practice of information behaviour research? As argued throughout this paper, detecting, processing, and using information is essential to the life activities of all living things. In information studies, we research the role and impact of information on humans in individual experience, including needing, seeking, finding, and encountering information, as well as in societies and ecosystems. This area of research is called information behaviour. The focus of this book is on the study of information in the interaction among humans, animals, and plants.

First, let us note the role of Gibson's and Barker's work in the field of psychology. Ecological psychology has run in parallel, as a sort of minor key, along with mainstream psychology for most of its existence (Segundo-Ortin & Raja, 2024, p. 27). It is as if the field of psychology recognises that there is a value in the metatheory but is unwilling to give up its mainline theories in favour of it, *nor* can the mainstream show that Gibson and Barker are wrong. Publishing from the ecological psychology perspective continues apace, the work continues to be cited, and the journal *Ecological Psychology* is in its 36th volume. Recent assessments of Gibson's work (Segundo-Ortin & Raja, 2024) and Barker's (Avram et al., 2024) attest to the vitality of their research paths and the remaining work that needs to be done in the area.

In a certain sense, awareness of the importance of the physical substrate of our information seeking has been there all along in library and information science, though it crops up in various, and usually incomplete, ways. In 1968, George Miller published an article in *American Documentation* (the initial name of the current *Journal of the Association for Information Science and Technology*) in which he argued for the importance of spatial dimensions of search. 'An aspect of the human use of information that has generally been overlooked in the automation of information services is the human tendency to locate information spatially' (Miller, 1968, p. 286).

Several major information behaviour researchers have included a strong ecological component in their writing. In 1989, I proposed a 'berrypicking' model of searching for information that emphasises the scattered process of gathering information on a topic, analogous to picking berries in a forest. (See also Savolainen's comparison of the berrypicking model and the 'information foraging' model, the latter drawing some attention in the human-computer interaction field, Savolainen, 2018.) See also Sandstrom, 1994. I built upon my 1989 work in a later paper analysing the impact of searching of the comparative *densities* on relevant resources across different territories (Bates, 2002).

Both physical and social constraints on information seeking are very evident in Elfreda Chatman's work on people operating in distinctive behaviour settings in prisons (1999), doing janitorial work (1990), and living in a retirement home (1992). Each milieu strongly influenced the people working or living within it.

Pettigrew brought us the concept of 'information ground,' namely, familiar locations where people feel comfortable sharing information (Pettigrew, 1999). Sonnenwald developed 'information horizons,' a term describing the range and limits of a person's information source world (Sonnenwald, 2005).

Despite the (often implicit) importance of place in these bodies of work, place has still not taken a very large role in the work either of Library and Information Science (LIS) scientists or of information system designers. I argued for an explicitly physical model of information searching in the design of search interfaces in a keynote speech at the Information Architecture Conference in 2018 (Bates, 2018b), saying that searching information on a screen should be done in the context of a visible path through a forest, with paths diverging to different geographical areas representing topical areas in the forest. If we spent tens of thousands of years as hunter-gatherers, why not model our search on that deeply ingrained physical experience? I find it personally astonishing that, over 50 years after Miller admonished us to pay more attention to physical location in our information system design, we have still not taken up a geographical model in any widely used information system.

Recently, however, the ecological level has been more thoroughly developed in LIS in the numerous studies of Annemarie Lloyd, who wrote of 'information landscapes,' often drawing on the language of Gibson in discussing informational affordances of particular situations and places (Lloyd, 2006, 2010). Savolainen conducted a detailed analysis of Lloyd's 'information landscapes' concept and its usage (2021).

Lloyd's concept of information landscape is a complex one.

> For example, when people enter a doctor's office they enter a physical location, which is imbued with meaning. When they enter into a relationship with their doctor they are entering into a space through which inter-subjective agreement about the meaning of their interaction is traded. Information and knowledge are exchanged at both levels creating the information landscape. This information landscape is socially constructed and emplaced within a larger health information environment, which is in turn composed of a wide range of domain specific and structural knowledges.
>
> *(Lloyd, 2014, p. 1037)*

For Lloyd, the information landscape is 'both a physical and metaphysical place.' It 'is constructed through engagement with different sites of knowledge and with varied sources of information that are embodied, social and textual' (Lloyd, 2014, p. 1037). Lloyd's study of ambulance officers beautifully exemplified these three broad types of information sources (Lloyd, 2009). The drivers received book learning in classes, social learning from their colleagues, and embodied learning from coming to read the physical state of their patients' bodies. More recent work by Lloyd concerns the resettlement experiences of migrants—the breakup of their old configuration of information sources and the establishment of a new configuration in the new country (2014). The importance of the destruction of older information landscapes is evident in her subjects' experience, as well as the difficulty of restructuring new information landscapes in their new home countries.

Finally, consciousness of the ecology of information-related behaviours has come strongly to the fore with the work of this volume's editor, Niloofar Sohljoo. 'While the information experienced by a dog or cat may differ from that of a human, a suitable starting point for investigating information lies in their shared socio-physical environment' (Solhjoo, Krtalić, & Goulding, 2024, p. 702). The physical ecology shared by humans and animals deserves far more attention than it has received to date, and Solhjoo is laying the foundation of this enlarged conception of information environments and information transfer in her research. See also Solhjoo (2024) and other works of hers in this volume.

Conclusion

In conclusion, in studying the movement and use of information among several species, it would seem to be a good idea to ensure that the physical, ecological layer is fully developed and incorporated into the research design. The behaviour setting and physical ecology are what humans and animals most closely and extensively share. This ecological component should constitute the foundation of a complete research design in many instances. In doing such research, one would:

- Describe the characteristics of the umwelts of the several species involved.
- Identify and characterise the physical ecology and its affordances for the several species involved.
- Identify and characterise the behaviour settings of significance to the research question(s) of the study.

Several of the other metatheoretical approaches listed in Figure 2.1 could be integrated fairly easily in a research design with the ecological psychology approach: Embodiment and materiality, affective, cognitive, user-centred design, evolutionary, socio-cognitive, and constructivism, in particular.

Information is fundamental to the lives of all living systems. Within our bodies—whether plant or animal—we have elaborate systems to enable the detection, storage, and use of information. It is only appropriate that we in the information studies field should research this fundamental and pervasive part of our lives. Further, given the importance of animals and plants in human lives, it is now time to extend our theory and research to the social, emotional, and operational roles of information in our interaction with other species.

References

Avram, C.M., Jones, A.E., Lucas, M.L., & Barrett, L. (2024). Reclaiming behaviour settings: Reviewing empirical applications of Barker's behaviour settings theory. *Philosophical Transactions of the Royal Society B*, 379(1910), 20230283. http://dx.doi.org/10.1098/rstb.2023.0283

Barker, R.G. (1968). *Ecological psychology: Concepts and methods for studying the environment of human behaviour*. Stanford, CA: Stanford University Press.

Barker, R.G., & Wright, H.F. (1955). *Midwest and its children*. New York: Harper & Row.

Bates, M.J. (2002). Speculations on browsing, directed searching, and linking in relation to the Bradford Distribution. In H. Bruce, R. Fidel, P. Ingwersen, & P. Vakkari (Eds.), *Emerging frameworks and methods: Proceedings of the fourth international conference on conceptions of library and information science (CoLIS4)* (pp. 137–150). Greenwood Village, CO: Libraries Unlimited. Also in: M. J. Bates (2016), *Information searching theory and practice, Selected works*. Berkeley, CA: Ketchikan Press, vol. 2, pp. 279–293.

Bates, M.J. (2005a). An introduction to metatheories, theories, and models. In K.E. Fisher, S. Erdelez, & L. McKechnie (Eds.), *Theories of information behaviour* (pp. 1–24). Medford, NJ: Information Today.

Bates, M.J. (2005b). Information and knowledge: An evolutionary framework for information science. *Information Research*, 10(4). http://www.informationr.net/ir/10-4/paper239
Bates, M.J. (2006). Fundamental forms of information. *Journal of the American Society for Information Science and Technology*, 57(8), 1033–1045.
Bates, M. J. (2007). What is browsing – really? A model drawing from behavioural science research. *Information Research*, 12(4), paper 330. http://InformationR.net/ir/12-4/paper330.html
Bates, M.J. (2018a). Concepts for the study of information embodiment. *Library Trends*, 66(3), 239–266.
Bates, M.J. (2018b). Neolithic information seeking: Designing information systems for our inner hunter-gatherer. (Keynote speech) Information Architecture Summit 2018, Chicago, IL: March 24, 2018. (Text of Speech) (Powerpoint slides) (Video).
Bates, M.J. (2022). A proto-paradigm for information science research. In *Proceedings of CoLIS, the 11th international conference on conceptions of library and information science*, Oslo, Norway, May 29–June 1, 2022. Information Research, 27 (Special issue), paper colis2201. http://informationR.net/ir/27-SpIssue/CoLIS2022/colis2201.html. https://doi.org/10.47989/colis2201
Chamovitz, D. (2012). *What a plant knows: A field guide to the senses*. New York: Farrar, Straus and Giroux.
Chatman, E.A. (1990). Alienation theory: Application of a conceptual framework to a study of information among janitors. *RQ*, 29, 355–368.
Chatman, E.A. (1992). *The information world of retired women*. Westport, CT: Greenwood Press.
Chatman, E.A. (1999). A theory of life in the round. *Journal of the American Society for Information Science*. 50(3), 207–217.
Gibson, J.J. (1979/2015). *The ecological approach to visual perception*. Classic ed. New York: Psychology Press.
Heft, H. (2001). *Ecological psychology in context: James Gibson, Roger Barker, and the legacy of William James' radical empiricism*. New York: Psychology Press.
Kovac, L. (2000). Fundamental principles of cognitive biology. *Evolution and Cognition*, 6(1), 51–69.
Kovac, L. (2007). Information and knowledge in biology: Time for reappraisal. *Plant Signaling & Behaviour*, 2(2), 65–73.
Lloyd, A. (2006). Information literacy landscapes: An emerging picture. *Journal of Documentation*, 62(5), 570–583.
Lloyd, A. (2009). Informing practice: Information experiences of ambulance officers in training and on-road practice. *Journal of Documentation*, 65(3), 396–419.
Lloyd, A. (2010). *Information literacy landscapes: Information literacy in education, workplace and everyday contexts*. Oxford: Chandos Publishing.
Lloyd, A. (2014). Stranger in a strange land; Enabling information resilience in resettlement landscapes. *Journal of Documentation*, 71(5), 1029–1042.
Lorenz, K. (1977). *Behind the mirror* (R. Taylor, Trans.). London: Methuen & Co.
Miller, G.A. (1968). Psychology and information. *American Documentation*, 19(3), 286–289.
Odling-Smee, F.J., Laland, K.N., & Feldman, M.W. (2003). *Niche construction: The neglected process in evolution*. Princeton, NJ: Princeton University Press.
Ofri, R. (2018). Vision in dogs and cats. *American Veterinarian*, 3(6). https://www.dvm360.com/view/vision-in-dogs-and-cats#
Pettigrew, K.E. (1999). Waiting for chiropody: Contextual results from an ethnographic study of the information behaviour among attendees at community clinics. *Information Processing & Management*, 35(6), 801–817.

Sandstrom, P.E. (1994). An optimal foraging approach to information seeking and use. *Library Quarterly*, 64(4), 414–449,

Savolainen, R. (2018). Berrypicking and information foraging: Comparison of two theoretical frameworks for studying exploratory search *Journal of Information Science*, 44(5), 580–593.

Savolainen, R. (2021). Information landscapes as contexts of information practices. *Journal of Librarianship and Information Science*, 53(4), 655–667.

Segundo-Ortin, M., & Raja, V. (2024). *Ecological psychology*. Cambridge Elements: Perception. Cambridge: Cambridge University Press.

Sheldrake, M. (2020). *Entangled life: How fungi make our worlds, change our minds, & shape our futures*. New York: Random House.

Solhjoo, N. (2024). Knowing within multispecies families: An information experience study. *Journal of Information Science*. https://doi.org/10.1177/01655515241268845

Solhjoo, N., Krtalić, M., & Goulding, A. (2024). The in-between: Information experience within human-companion animal living environments. *Journal of Documentation*, 80(4), 701–729.

Sonnenwald, D.H. (2005). Information horizons. In K.E. Fisher, S. Erdelez, & L. McKechnie (Eds.), *Theories of information behaviour* (pp. 191–197). Medford, NJ: Information Today.

Uexküll, J. von. (1934/2010). *A foray into the worlds of animals and humans, with a theory of meaning*. Translated by J.D. O'Neil. Minneapolis: University of Minnesota Press.

University of Adelaide, Faculty of Sciences, Engineering, and Technology. How far away can dogs smell and hear? Posted June 9, 2020. https://set.adelaide.edu.au/news/list/2020/06/09/how-far-away-can-dogs-smell-and-hear#:~:text=How%20far%20dogs%20can%20smell,not%20the%20only%20great%20smellers (Accessed 11/23/2024)

West Suburban Humane Society. (2024). *What do pets see?* https://www.wshs-dg.org/resource-center/ask-the-vet-pet-watch/132-ask-the-vet/479-what-do-pets-see#:~:text=Dogs%20and%20cats%20have%20binocular,degrees%20compared%20to%2060%20degrees (accessed 11/23/2024)

Wiener, N. (1961). *Cybernetics: Or control and communication in the animal and the machine* (2nd ed.). Cambridge, MA: The MIT Press.

Wohlleben, P. (2017). *The hidden life of trees*. Glasgow: William Collins.

Yong, E. (2022). *An immense world: How animal senses reveal the hidden realms around us*. London: Penguin Random House.

3

INFORMATION EXPERIENCE FOR THE EXPANDING CIRCLE

Insights from Animal–Computer Interaction and More-Than-Human–Computer Interaction

Tim Gorichanaz

Introduction

Every morning, my cat Toaster asks for a few butt-pats or scratches. Occasionally, he may ask for brushing. He's a 12-year-old grey Siberian who leaves his long hair everywhere.

A few months back, I bought a set of talking buttons for Toaster. You've likely seen videos of them online. The set I bought is a modular system of foam tiles and programmable buttons from the company FluentPet. With each button, you can record an audio clip to play when the button is pushed, and you can place the button wherever you'd like on the foam tile (and place that tile wherever you'd like in your house), and then train your pet to understand the association between the button and an action. We started Toaster with three buttons: Brush, scratches, and butt-pats (butt-pats being his particular passion, perhaps unlike most cats). Within a month, Toaster learnt all three buttons. Now he uses them to very deliberately ask for what he wants. And sometimes he just sits in front of the pad of buttons, perhaps in the way that I sometimes sit idly at the computer, daydreaming.

This device has enabled a new form of information exchange between my cat and his humans, and it's given him a new form of agency. That is technology design at its best: Enabling.

That such a device exists makes me wonder about the process by which it was designed—how the creators arrived at these particular details—and also about the informational world of my cat. What information is present in his life? What might he want to communicate? Does his newfound agency over the flows of information in the world make his life better?

DOI: 10.4324/9781003583424-4

Essentially, these are questions of information experience, but not human information experience. And in the context of a computerised device, they are questions of computer interaction, but not human–computer interaction. Rather, they gesture towards new horizons for existing research in information and computing.

In information science, the research field of information behaviour (IB) has been expanding in scope in recent decades, examining information in more contexts and considering the informative qualities of more phenomena than previously recognised (Bates, 2006, 2017). The emergence of information experience and multispecies information is further evidence of this expansion.

Meanwhile, the field of human–computer interaction (HCI) has been expanding similarly, to the point that the fields of IB and HCI are beginning to converge in certain respects (Gorichanaz & Venkatagiri, 2022). Identifying shared interests between IB and HCI is an opportunity for mutual learning between these two fields, which are, to date, still largely siloed apart in distinct academic communities.

In that spirit, this chapter takes a cue from animal-centred and multispecies work from HCI to provide lessons for the incipient animal turn within information science. This chapter discusses the methods and ways of knowing established within animal–computer interaction (ACI), and it illustrates how animal-centred research provides metatheoretical value to information experience, information behaviour and information science more generally.

Information Behaviour, Information Experience, and Human–Computer Interaction

Before proceeding, a terminological note is in order. My research has centred on information experience (Gorichanaz, 2020), which examines people's in-the-moment engagement with information from their own perspective (Gorichanaz, 2017b, 2019a). Information experience can be understood as an object of study, as well as a research area (which examines information experiences as its object of study) (Partridge & Yates, 2014). One benefit of the information experience perspective is that it shows the informativeness of many phenomena that have not traditionally been considered as information in the field of information science (Gorichanaz, 2023b).

As a research area, I consider information experience to sit within the umbrella of IB. In turn, I favour a broad definition of IB (Huvila & Gorichanaz, 2024) to include all the ways in which humans need, seek, use, create, destroy… and experience information, regarded from any theoretical standpoint. Other scholars prefer to see information experience as separate from IB, which they may define more narrowly, sometimes seeing information experience more as a matter of information literacy (Gorichanaz, 2020).

While IB looks at the ways people interact with information, the field of HCI looks at the ways people interact with computers. Given that much of the information people interact with today is mediated by computers, IB and HCI today have significant overlap, as Venkatagiri and I have documented (Gorichanaz & Venkatagiri, 2022), though, as mentioned above, these research fields are still separated in practice. I believe the perspective of information experience can help demonstrate the synergies between these fields and enable mutual learning (Gorichanaz, 2020), which is the spirit in which I offer this chapter.

An Animal Turn in Information Science

Information science is ultimately about people (Furner, 2004), like virtually all fields of inquiry across the sciences and humanities. This is, perhaps, evidence of our broader anthropocentrism.

But gradually, especially over the past 50 years, humans have been acknowledging the inherent worth of other species and even non-living entities we share the world with, as well as how their existence enables and enriches our own. Even if you only care about human beings, you must acknowledge that we need oxygen to breathe and food to eat and thus depend on myriad interlocking ecological cascades for our survival.

Recent decades have witnessed rises in animal rights and environmentalism, among other currents, as well as a turn towards the animal in many fields of academic research (Andersson Cederholm et al., 2014; Ritvo, 2007). In HCI, Mancini (2011) observed that technology designed for animals was missing the animals' perspective, counter to the prevailing philosophy of user-centred design. But as Toaster shows us with his buttons, humans are not the only users we should imagine. In her now-famous manifesto, Mancini called for further research in ACI as a way to improve animal welfare, address sustainability, and enrich the field of HCI.

In information science, Hartel (2019) observed a number of the field's turns throughout history, wondering if an animal turn may be on the horizon. If so, Solhjoo's (2024) doctoral thesis is a landmark in that regard. But it's worth noting other precedents. In the mid-20th century, French writers wondered about the documentary properties of animals such as gorillas and antelopes (Briet, 1951/2006; Pagès et al., 1948/2021), leading to more recent commentary and theorisation about animal rights (Day, 2018). Other authors, such as myself, have speculated about whether definitions of information and evidence must be confined to human beings (Gorichanaz, 2015). And scholars such as Grenersen (2016) have commented on the human–animal relationship when it comes to information, as evident in the case of the Sámi people and reindeer herds in the Arctic.

As the animal turn continues to unfold across academia, it is likely that such work in information science will continue to develop. This chapter, then, may serve as a metatheoretical and methodological guide to that end.

Research Approaches in Animal–Computer Interaction

The most common data-gathering techniques in information behaviour and HCI include interviews and surveys (Huvila & Gorichanaz, 2024; Lazar et al., 2017). Clearly, these techniques are not suitable for research with animals. They cannot speak to us or fill out questionnaires. What's more, we humans cannot necessarily infer animals' affective states by their behavioural responses (Hirskyj-Douglas et al., 2018; Ritvo & Allison, 2014)—for example, chimpanzees smile when they are afraid (Lietman, 2022).

Given these methodological challenges, Lawson et al. (2016) noted that most ACI approaches to technology for animals have been technology-driven or human-centred rather than animal-centred; that is, the resulting products, such as feeding cameras and toys, were more for human entertainment than for the animals themselves. Lawson worries that such approaches risk animals becoming 'usees' (Baumer, 2015), or inadvertent users who did not consent to using a system that subversively exploits them. Researchers in animal–computer interaction have recognised that new methods and methodological approaches are needed to successfully do animal-centred work (Hirskyj-Douglas et al., 2018; Lawson et al., 2016; Ritvo & Allison, 2014).

The two approaches now most common and thought to be most successful in ACI are phenomenology and speculative design. Phenomenology is an interpretative approach that attempts, in the case of ACI, to help people imagine what the world is like for a particular animal, given its different biological capacities and life (Lestel et al., 2014). An example of such work is that of Hirskyj-Douglas and Read (2014), which considers dogs on their own terms to examine what technology that is truly dog-centred would be like. Phenomenological methods have been developed and supported in HCI outside ACI (Gunkel, 2018), and along with phenomenography, phenomenology is a leading approach in information experience and other areas of information science (Dabengwa et al., 2023; Gorichanaz, 2020). This suggests that the perspective of information experience is poised to support multispecies work in information science and computer science, such as has been demonstrated by Solhjoo (2024).

Lestel et al. (2014) offer some guidance for conducting phenomenological research with animals: Study the animal on its own terms (not, e.g., asking it to complete human-concocted tasks), look for what is shared between animals and humans and what is not, and observe that animal intelligence comes in multiple forms (including cooperation and cunning). In terms of specific methods appropriate to animal-centred research, phenomenology may involve observing animals, interviewing the people in the animals' lives, studying the relevant animal science, and conducting phenomenological writing (see van Manen, 2014).

In addition to its phenomenological approach, other aspects of existing information experience theory are well-positioned to support further research in animal–computer interaction and multispecies information science: The first-person perspective (Gorichanaz, 2019a), the concepts of a priori and a posteriori information (Yu & Liu, 2022), and the notions of self, understanding and meaning (Gorichanaz, 2020)—these are all relevant to the study of animal interactions as well as human ones.

The next approach that has become standard in ACI is speculative design, which uses design methods to imagine (and help others imagine) possible futures. Whereas standard design practice is meant to create a solution for the marketplace, speculative design more commonly is meant to raise questions or provoke reflection by inviting people to confront an artefact from a possible future or alternative present (Dunne & Raby, 2013). Speculative design is commonly employed in ACI in tandem with phenomenology (if only implicitly), in the sense that effective speculation requires openness to others' lifeworlds. To cite one example from ACI, Lawson et al. (2016) observe that most technology for dogs is human-centred rather than dog-centred, and they sketch a 'dog internet' as an exercise to imagine what dogs in and of themselves might want, such as alerts to know when their owner is coming home soon and interfaces that engage the dog's sense of smell. Specific methods for speculative design in ACI have included design workshops (Nijs, Laki, Houlstan, Slizewicz, & Laureyssens, 2020) and fiction exercises (Rodgers et al., 2023), but speculative design involves a range of methods developed in design research and HCI that have yet to be applied in ACI, such as games. Moreover, speculative design has yet to be applied within information behaviour, but it may be a fruitful approach to shed new light on how people need, seek, use and experience information.

Considering the ethical and regulatory aspects of research is familiar when the research participants are human beings. It is just as vital to consider the ethics of any given research project involving animals. As Mancini (2017) writes, most frameworks for research ethics do consider animal welfare, but only as a concern secondary to human scientific goals. Mancini proposes an ethical framework for animal research that treats animals as individuals rather than research instruments. Because consent is a key issue in research ethics and a major challenge when it comes to animals, the framework includes mediated consent, in which a fiduciary agent brokers consent on the animal's behalf, and contingent consent, allowing the animal to make relevant choices on an ongoing basis and to withdraw from the research altogether at will.

Ways of Knowing and Research Contributions in Animal-Centred Research

The philosopher Ludwig Wittgenstein quipped, 'If a lion could talk, we wouldn't be able to understand it' (Wittgenstein, 1953, p. 235). That is, a lion's world and concerns are so different from ours that the language barrier is not the biggest obstacle to understanding each other. Another philosopher, Thomas Nagel (1974), similarly pointed out how hopeless it is to imagine what it's like to be a bat.

These notions overstate the case. Empathy and imagination are basic human skills. Any time we are gripped by a good story, we are imagining the lives of others, moving towards understanding those experiences. Nonhuman animals may be more different from us than other humans, but still we share much in common: The need for food and shelter and movement, the Earth as our home, and so on.

Still, it is worth reflecting: Given the challenges involved in researching with animals, what kinds of research contributions are possible, and how can we trust the knowledge generated from such research?

First, we must recognise that all knowledge generated in animal-centred work is situated. Indeed, situatedness is characteristic of interpretive research generally (Williamson & Johanson, 2018), including that in information experience. This means that the results from such research cannot be statistically generalised to broader populations, but it must be applied analytically on a case-by-case basis, taking into account the context of the originating study and the target context. As Lestel et al. (2014) write, the goal with animal-centred research is not to create a standalone model of a phenomenon (which is impossible), but to bring knowledge about animal life into contact with human life so that we humans can live better together with animals. The approach that Lestel advocates, named bi-constructivism:

> highlights that invention and interpretation are part of both animal activity and ethological research into it. Rather than trying to bracket the humans out of the picture...bi-constructivism pays heed to precisely this interface as a crucial aspect of the phenomenological and social setting and a real and significant element of animals' being and ecology.

This observation leads to the second key consideration about knowledge in animal-centred research: Its pragmatic or heuristic nature, which can be called a 'tinkering' approach to knowledge generation. This is characteristic of design research and HCI more generally. In design, knowledge is generated through making, which involves hands-on exploration, iteration and reflection (Clarke, 2018). (As a note, Clarke argues that such an epistemology is also relevant and applicable to information science more broadly).

The tinkering approach is well-suited to addressing the needs of nonhuman animals, even if it cannot objectively and entirely uncover their experiences. We may never know what it's like to be a bat, as Nagel (1974) says, but through trial and error, we can see what sorts of interactions bats carry out and what sorts of products bats prefer. That process generates knowledge. The pragmatic, do-what-works approach allows researchers to find solutions without wading into the thornier questions of ultimate truth. As Lestel et al. (2014, p. 130) write, 'It shifts the frame from the Aristotelian question 'What is an animal?' to the Spinozist inquiry "What can it do?"' This is also characteristic of Luciano Floridi's philosophy of information, which puts the focus of ontology and epistemology on interactivity (Floridi, 2019). Similarly, Coeckelbergh (2011) and Gunkel (2018) emphasise that in phenomenological research with nonhuman entities, what matters most is not factual correctness but rather communicative behaviour and efficacy. To that end, perhaps the question being asked in animal-centred research is not just 'What can it do?' but 'What does it want to do?' as one reviewer of this chapter suggested.

Wang et al. (2024) offer an example of a study that illustrates situated knowledge and the tinkering approach. They present research on designing toys and entertainment devices for lemurs in a zoo. The key question was whether a single-sensory device or a multi-sensory one would be more engaging to the lemurs. They observed that to date, most enrichment devices for lemurs only provided a single-sensory experience (e.g., animations on a screen), but lemurs engage with their environments with multiple senses (particularly smell). To answer their research question, Wang et al. deployed in a lemur enclosure a device that could show video with accompanying smell and audio; several different combinations of single-sensory and multi-sensory output were tested over a seven-week period. The specifics of the device were informed by research on the biological capacities of lemurs (what colours they see, what aspects of their environment are interesting, etc.). In the end, the researchers found that lemurs engaged longer with the device when it had multi-sensory output, and they spent more time in the space where the device was present. One result from this study was the device itself, but the study also furnished knowledge about the engagement preferences and behaviour of lemurs—heuristic knowledge that can be applied towards further design experiments.

Thinking again about my cat, Toaster, and his buttons, not only did the company FluentPet use a tinkering approach to design and optimise their product (and they continue to do so), but I also used such an approach when setting up the device and introducing it into Toaster's life. The modular tiles and programmable buttons inherently encourage tinkering. The result has been knowledge that is situated and useful within my family. For instance, while FluentPet's marketing shows animals tapping the buttons with their paws, Toaster prefers to press them with his nose. The tinkering and learning continue: I'm noticing that Toaster seldom (okay, more like never) pushes the button for brushing, though, as a long-haired cat, he needs to be brushed every day to prevent his hair from matting. Yet I can be the one to push the button for brushing when it's time so that he better understands my intentions. This knowledge applies to our life together, not to all cats or even all Siberian cats, but this knowledge is still valuable in that it improves our getting along together. But there is value beyond our family. For example, my findings from Toaster's interactions with his buttons could suggest to others some possibilities to try with their own cats. And I believe there is value even beyond this, which I explore in the next section.

The Value of Animal Research in the Information and Computer Sciences

What good is animal-centred research? What place could it have in existing research fields? For information science in particular, is there any value that multispecies work may have?

For one, it may be that such work is simply interesting. I follow Buckland (2000) in my conviction that there is space even in professional fields such as information science for work that flows like a labour of love in the spirit of the liberal arts. Even if there were no direct application of such work, and even if no tacit value ever came out, it would still be worthwhile for researchers to examine what is interesting. It is like the philosopher Martin Heidegger wrote of philosophy: 'even if we can't do anything with it, may not philosophy in the end do something with us, provided that we engage ourselves with it?' (Heidegger, 2014, p. 13).

But I do believe that there is also immediate practical relevance of multispecies work in information science. To begin, Mancini (2013) writes on the value of ACI for the field of HCI. As she explains, ACI contributes towards HCI's stated aims of sustainability, justice, and broadening participation in computing. It also strengthens HCI as a discipline by providing a testbed for methodological experimentation; insights from working with animals may be applied to humans, perhaps most obviously infants but, others as well. And finally, ACI makes HCI a more complete research field, and indeed, it could be argued that ACI actually encompasses HCI, seeing as humans are a subset of the category of animals. These observations also apply to multispecies information science regarding the value it provides to information science writ large. To give one example of the kind of metatheoretical contribution that multispecies information science offers the broader field, it is already evident that findings from multispecies work have helped clarify a conceptual issue regarding the concept of information, which has been of great interest to me over the years: The distinction between information on the surface and below the surface. Information on the surface includes what we typically think of as information: Epistemic content. Information below the surface is more ineffable: Our ways of being are influenced by information flows.

Our information-laden interactions with animals primarily highlight information below the surface because we cannot communicate with animals in the same way we do with humans. As a result, as Solhjoo (2024) writes, information with animals is characterised by loving, living, and learning—these are not facts per se, but changes in our way of being.

Previous work has gestured towards this distinction, and future work should further clarify it. For instance, in a paper focusing on the concept of understanding, I distinguished between ontic understanding and ontological understanding (Gorichanaz, 2017a). Later, Latham and I wrote of 'contemplative aims' for information, discussing the ways in which becoming informed can shape our attention, compassion, unity, and so on (Gorichanaz & Latham, 2019). More recently, I have been investigating how information may contribute to virtue development, which operates similarly (Gorichanaz, 2023a). The distinction between information above and below the surface is also evident in Ruthven's work on information sculpting (Ruthven, 2022). Likely other work has addressed the issue using different terminology as well; as I mentioned, future work would do well to review and clarify this terminology. Here, I only wish to point out that multispecies work in information science can play a large role in that clarification.

Multispecies Research beyond Animals

Peter Singer's (1981) work in *The Expanding Circle* was a landmark for animal rights in the post-industrial West. In that book, Singer argues that humanity's circle of moral concern has been expanding throughout history from close kin and tribes to communities and nations and now beyond humanity to include animals. The expansion has not stopped there.

On a parallel track, the notion of rights of nature—that ecosystems, nonhuman species, landforms, etc., should have moral and legal standing—has also emerged. Rights of nature are being enshrined in law in more and more countries.[1] Hence, certain rivers and mountains are now protected and may file suit if violated. Recognition is dawning that we humans are not here to subjugate the rest of the world but rather to shepherd it.

Contemporary philosophical grounding for animal rights and the rights of nature comes from many corners, but most relevant to information science, we can look to the work of Luciano Floridi (2013). Floridi's information ontology and concept of the ontic trust show how all entities—from laptops to ladyslippers—are deserving of moral respect, roughly in accordance with their complexity (Gorichanaz, 2019b). This perspective is already beginning to be explored in HCI. It should be noted that even if such a perspective seems novel in contemporary research and legislation, several Indigenous knowledge systems have long held it (Benoit et al., 2024; Raven et al., 2021). In that light, future work could learn from such traditions as it moves forward.

Looking beyond ACI, a budding 'more-than-human' mindset in HCI strives to include the participation of entities in addition to nonhuman animals that have been excluded from consideration in HCI and technology design more generally (Wakkary, 2021).

This type of work has been bubbling up in several socio-technical fields of study in recent decades, inspired, for instance, by Bruno Latour's actor–network theory, but it has become increasingly urgent given the recent proliferation of artificial intelligence technology over the past decade (Eriksson et al., 2024). In the area of more-than-human HCI (sometimes called 'posthuman'), similar research approaches are being taken to what was discussed above for animal-centred work. Phenomenology as a research approach, for instance, is valued for studying human–robot interaction, as it enables researchers to grapple with the problem of other minds and explore how the objective informs the relational and subjective (Behrenshausen, 2007; Bisconti et al., 2024; Coeckelbergh, 2011; Pasek, 2015; Soma, 2024; Viik, 2020). In terms of the immediate value of such research, studies of non-animate sensory systems may also be applied fruitfully to enable or improve the lives of humans, such as machine learning systems to help substitute for vision loss (Buccella & Springle, 2023).

One interesting study in the realm of more-than-human HCI that can serve as an example centre on an AI-infused lamp called Message Ritual (Rajcic & McCormack, 2023). A person places the lamp somewhere in their house, where it constantly listens to the sounds and conversations that occur within its range. Each morning, the lamp sends an SMS to its owners sharing a poem of sorts that remixes key phrases and topics from what the lamp heard the previous day. Rajcic and McCormack (2023) invited four households to live with Message Ritual over a two-week period and then participate in an interview about their experiences. One of the key findings was that participants found themselves wanting to spend time in the lamp's vicinity to have certain conversations—even people who had been seeking to reduce their screen time, showing that not all digital technologies are equal, and indeed that technology can help us connect with each other rather than just obstruct genuine human connection. Another interesting finding was that participants experienced the lamp as a person, whether a housemate or guest, and treated it as a participant in their conversations.

This type of work may provide inspiration for a broader multispecies information science, one that is more-than-human and more holistic.

To speak of information experience specifically, one path for future work in this vein lies in the concept of the self. How might a human–animal and more-than-human perspective extend, enrich, and challenge existing understandings of the self? If a human self is constituted by information and documents, as I have written about in my prior work (Gorichanaz, 2020), what sort of information and documents might constitute an animal self? Considering how the human self can be extended by technologies such as eyeglasses, prosthetic limbs, and smartphones, how might we understand the boundaries of the animal self? And reflecting upon the tight coupling between animals and their humans in the home, how might we understand the notion of self in the move from 'me' to 'us'?

Conclusion

When my cat Toaster pushes his buttons, he is participating in the expansion of humans directing our technology beyond our own species, to our mutual benefit. This chapter has suggested that information science can draw inspiration and guidance from research currents in animal–computer interaction and more-than-human–computer interaction, such as those that have led to products like Toaster's buttons.

As I have pointed out, beyond its inherent value, such work may illuminate aspects of information science that have not yet been fully appreciated, such as the distinction between information above and below the surface. Certainly, more discoveries are waiting. For now, Toaster is calling for more butt-pats.

Note

1 See https://en.wikipedia.org/wiki/Rights_of_nature_law (Wikipedia. (n.d.). Rights of nature law. In Wikipedia. Retrieved October 29, 2025, from https://en.wikipedia.org/wiki/Rights_of_nature_law).

References

Andersson Cederholm, E., Björck, A., Jennbert, K., & Lönngren, A.-S. (2014). *Exploring the animal turn: Human-animal relations in science, society and culture.* Lund, Sweden: Pufendorf Institute for Advanced Scholars, Lund University.

Bates, M. J. (2006). Fundamental forms of information. *Journal of the American Society for Information Science and Technology*, 57 (8), 1033–1045. Retrieved from https://onlinelibrary.wiley.com/doi/abs/10.1002/asi.20369. https://doi.org/10.1002/asi.20369

Bates, M. J. (2017). Information behavior. In J. D. McDonald & M. Levine-Clark (Eds.), *Encyclopedia of library and information sciences* (4th ed.). Boca Raton, FL: CRC Press. https://doi.org/10.1081/E-ELIS3-120043263

Baumer, E. P. (2015). Reflective informatics: Conceptual dimensions for designing technologies of reflection. In *Proceedings of the 33rd annual ACM conference on human factors in computing systems* (pp. 585–594). New York: ACM. Retrieved from http://doi.org/10.1145/2702123.2702234

Behrenshausen, B. G. (2007). *Touching is good: An eidetic phenomenology of interface, interobjectivity, and interaction in Nintendo's "Animal Crossing: Wild World"* (Master's thesis, University of Maine, Orono, ME, USA). University of Maine Digital Commons. https://digitalcommons.library.umaine.edu/etd/277/

Benoit, J., Penn-Clarke, C. R., Rust, R., Groenewald, D. P., Vickers-Rich, P., & Helm, C. W. (2024). Indigenous knowledge of palaeontology in Africa. *Geological Society, London, Special Publications*, 543 (1), 357–370. Retrieved from http://doi.org/10.1144/SP543-2022-236

Bisconti, P., McIntyre, A., & Russo, F. (2024). Synthetic socio-technical systems: Poiêsis as meaning making. *Philosophy & Technology*, 37 (3), 94.
Briet, S. (2006). What is documentation? In R. E. Day, L. Martinet, & H. G. B. Anghe-Lescu (Eds. & Trans.), *What is documentation? English translation of the classic French text* (pp. 9–46). Lanham, MD: Scarecrow Press. (Original work published 1951).
Buccella, A., & Springle, A. A. (2023). Phenomenology: Whats AI got to do with it? *Phenomenology and the Cognitive Sciences*, 22 (3), 621–636.
Buckland, M. K. (2000). *The academic heritage of library and information science: Resources and opportunities*. San Antonio, TX: Plenary session presented at the Association for Library and Information Science Education Annual Meeting. Retrieved from http://people.ischool.berkeley.edu/~buckland/alise00.html
Clarke, R. I. (2018). Toward a design epistemology for librarianship. *The Library Quarterly*, 88 (1), 41–59.
Coeckelbergh, M. (2011). Humans, animals, and robots: A phenomenological approach to human-robot relations. *International Journal of Social Robotics*, 3, 197–204.
Dabengwa, I. M., Young, S., & Ngulube, P. (2023). Rigour in phenomenological and phenomenography studies: A scoping review of library and information science research. *Library & Information Science Research*, 45 (1), 101219.
Day, R. E. (2018). Auto-documentality as rights and powers. *Proceedings from the Document Academy*, 5 (2), 3.
Dunne, A., & Raby, F. (2013). *Speculative everything: Design, fiction, and social dreaming* (2nd ed.). Cambridge, MA: The MIT Press.
Eriksson, E., Yoo, D., Bekker, T., & Nilsson, E. M. (2024). More-than-human perspectives in human-computer interaction research: A scoping review. In *Proceedings of the 13th Nordic conference on human-computer interaction*. New York: Association for Computing Machinery. Retrieved from https://doi.org/10.1145/3679318.3685408
Floridi, L. (2013). *The ethics of information*. Oxford: Oxford University Press.
Floridi, L. (2019). *The logic of information*. Oxford: Oxford University Press.
Furner, J. (2004). Information studies without information. *Library Trends*, 52 (3), 427–446.
Gorichanaz, T. (2015). For every document, a person: A co-created view of documents. *Proceedings from the Document Academy*, 2 (1), 9.
Gorichanaz, T. (2017a). Applied epistemology and understanding in information studies. *Information Research: An International Electronic Journal*, 22 (4), n4.
Gorichanaz, T. (2017b). Information and experience, a dialogue. *Journal of Documentation*, 73 (3), 500–508.
Gorichanaz, T. (2019a). A first-person theory of documentation. *Journal of Documentation*, 75 (1), 190–212.
Gorichanaz, T. (2019b). The self and the ontic trust: toward technologies of care and meaning. *Journal of Information, Communication and Ethics in Society*, 17(4), 395–408. https://doi.org/10.1108/JICES-09-2018-0073
Gorichanaz, T. (2020). *Information experience in theory and design*. Bingley: Emerald.
Gorichanaz, T. (2023a). A compass for what matters: Applying virtue ethics to information behavior. *Open Information Science*, 7 (1), 20220151.
Gorichanaz, T. (2023b). On the two conceptualizations of information experience as an object of study: A response to Yu and Liu. *Journal of Documentation*, 79 (3), 635–640.
Gorichanaz, T., & Latham, K. F. (2019). Contemplative aims for information. *Information Research*, 24(3). Retrieved from http://www.informationr.net/ir/24-3/paper836.html
Gorichanaz, T., & Venkatagiri, S. (2022). The expanding circles of information behavior and human–computer interaction. *Journal of Librarianship and Information Science*, 54 (3), 389–403. Retrieved from https://doi.org/10.1177/09610006211015782

Grenersen, G. (2016). Documentation, information and the animal connection. *Proceedings from the Document Academy*, 3 (2), 15.
Gunkel, D. J. (2018). The relational turn: Third wave HCI and phenomenology. In M. Filimowicz & V. Tzankova (Eds.), *New directions in third wave human-computer interaction: Volume 1 – Technologies* (pp. 11–24). Cham, Switzerland: Springer. https://doi.org/10.1007/978-3-319-73356-2_2
Hartel, J. (2019). Turn, turn, turn. *Information Research*, 24 (4), colis1901. Retrieved from http://informationr.net/ir/24-4/colis/colis1901.html
Heidegger, M. (2014). *Introduction to metaphysics* (2nd ed., G. Fried & R. Polt, Trans.). New Haven, CT: Yale University Press.
Hirskyj-Douglas, I., Pons, P., Read, J. C., & Jaen, J. (2018). Seven years after the manifesto: Literature review and research directions for technologies in animal computer interaction. *Multimodal Technologies and Interaction*, 2 (2). Retrieved from https://www.mdpi.com/2414-4088/2/2/30. https://doi.org/10.3390/mti2020030
Hirskyj-Douglas, I., & Read, J. C. (2014). Who is really in the center of dog computer design? In *Proceedings of the 2014 workshops on advances in computer entertainment conference*. New York: Association for Computing Machinery. Retrieved from https://doi.org/10.1145/2693787.2693793
Huvila, I., & Gorichanaz, T. (2024). Trends in information behavior research, 20162022: An annual review of information science and technology paper. *Journal of the Association for Information Science and Technology*. Retrieved from https://doi.org/10.1002/asi.24943
Lawson, S., Kirman, B., & Linehan, C. (2016, June). Power, participation, and the dog internet. *Interactions*, 23 (4), 3741. Retrieved from https://doi.org/10.1145/2942442
Lazar, J., Feng, J. H., & Hochheiser, H. (2017). *Research methods in human-computer interaction*. Burlington, MA: Morgan Kaufmann.
Lestel, D., Bussolini, J., & Chrulew, M. (2014). The phenomenology of animal life. *Environmental Humanities*, 5 (1), 125–148.
Lietman, A. (2022). *Chimp faces: Noticing & reading facial expressions*. Retrieved from https://projectchimps.org/reading_facial_expressions/
Mancini, C. (2011, July). Animal-computer interaction: A manifesto. *Interactions*, 18 (4), 6973. Retrieved from https://doi.org/10.1145/1978822.1978836
Mancini, C. (2013). Animal-computer interaction (ACI): Changing perspective on HCI, participation and sustainability. In *Chi '13 extended abstracts on human factors in computing systems* (p. 22272236). New York: Association for Computing Machinery. Retrieved from https://doi.org/10.1145/2468356.2468744
Mancini, C. (2017). Towards an animal-centred ethics for animal computer interaction. *International Journal of Human-Computer Studies*, 98, 221–233. Retrieved from https://www.sciencedirect.com/science/article/pii/S1071581916300180. https://doi.org/10.1016/j.ijhcs.2016.04.008
Nagel, T. (1974). What is it like to be a bat? *The Philosophical Review*, 83 (4), 435–450.
Nijs, G., Laki, G., Houlstan, R., Slizewicz, G., & Laureyssens, T. (2020). Fostering more-than-human imaginaries: Introducing DIY speculative fabulation in civic HCI. In *Proceedings of the 11th Nordic conference on human-computer interaction: Shaping experiences, shaping society*. New York: Association for Computing Machinery. https://doi.org/10.1145/3419249.3420147
Pagès, R., et al. (2021). Documentary transformations and cultural context. *Proceedings from the Document Academy*, 8 (1), 2. (Original work published 1948)
Partridge, H., & Yates, C. (2014). Researching information experience: Object and domain. In C. Bruce, K. Davis, H. Hughes, H. Partridge, & I. Stoodley (Eds.), *Information experience: Approaches to theory and practice* (pp. 19–32). Bingley: Emerald.

Pasek, A. (2015). The problem of nonhuman phenomenology: Or, what is it like to be a kinect? *In Visible Culture*, (22) Retrieved from https://ezproxy.csu.edu.au/login?url=https://www.proquest.com/scholarly-journals/problem-nonhuman-phenomenology-what-is-like-be/docview/1771028702/se-2

Rajcic, N., & McCormack, J. (2023). Message ritual: A posthuman account of living with lamp. In *Proceedings of the 2023 CHI conference on human factors in computing systems*. New York: Association for Computing Machinery. Retrieved from https://doi.org/10.1145/3544548.3581363

Raven, M., Robinson, D., & Hunter, J. (2021). The emu: More-than-human and more-than-animal geographies. *Antipode*, 53 (5), 1526–1545. Retrieved from https://doi.org/10.1111/anti.12736

Ritvo, H. (2007, 10). on the animal turn. *Daedalus*, 136 (4), 118–122. Retrieved from https://doi.org/10.1162/daed.2007.136.4.118

Ritvo, S. E., & Allison, R. S. (2014). Challenges related to nonhuman animal-computer interaction: Usability and 'liking'. In *Proceedings of the 2014 workshops on advances in computer entertainment conference*. New York: Association for Computing Machinery. Retrieved from https://doi.org/10.1145/2693787.2693795

Rodgers, S., Vella, K., Ploderer, B., & Brereton, M. (2023). Nature fictions: Designing for a sustainable future through nature relations. In *Proceedings of the 2023 ACM designing interactive systems conference* (p. 15661580). New York: Association for Computing Machinery. Retrieved from https://doi.org/10.1145/3563657.3595981

Ruthven, I. (2022). *Dealing with change through information sculpting*. Bingley: Emerald.

Singer, P. (1981). *The expanding circle: Ethics and sociobiology*. Princeton, NJ: Princeton University Press.

Solhjoo, N. (2024). Information experience within multispecies families: Understanding shared life with companion cats and dogs (Doctoral dissertation, Victoria University of Wellington). Open Access Te Herenga Waka–Victoria University of Wellington. https://openaccess.wgtn.ac.nz/articles/journal_contribution/The_In-between_Information_Experience_within_Human-Companion_Animal_Living/24986298

Soma, R. (2024). *Does a robot lawnmower know anything about grass? A phenomenological gaze on human–robot interactions* (Doctoral thesis, University of Oslo, Oslo, Norway). University of Oslo.

Van Manen, M. (2014). *Phenomenology of practice: Meaning-giving methods in phenomenological research and writing*. Walnut Creek, CA: Left Coast Press.

Viik, T. (2020). Falling in love with robots: a phenomenological study of experiencing technological alterities. *Paladyn, Journal of Behavioral Robotics*, 11 (1), 52–65. Retrieved 02/12/2024, from https://doi.org/10.1515/pjbr-2020-0005

Wakkary, R. (2021). *Things we could design: For more than human-centered worlds*.Cambridge, MA: MIT Press.

Wang, J., Brewster, S. A., & Hirskyj-Douglas, I. (2024). Investigating Lemurs sensory modality in technologies: How do lemurs engage with single and multimodal stimuli in zoos? In *Extended abstracts of the CHI conference on human factors in computing systems*. New York: Association for Computing Machinery. Retrieved from https://doi.org/10.1145/3613905.3651056

Williamson, K., & Johanson, G. (Eds.) (2018). *Research methods: Information, systems, and contexts* (2nd ed.). Sawston: Chandos Publishing.

Wittgenstein, L. (1953). *Philosophical investigations* (G. Anscombe & R. Rhees, Eds. & G. Anscombe, Trans.). Oxford: Blackwell.

Yu, L., & Liu, Y. (2022). Information experience as an object of LIS research: a definition based on concept analysis. *Journal of Documentation*, 78 (6), 1487–1508.

4

THE ANIMAL'S IMPORTANCE FOR NEO-DOCUMENTATION

A 'Copernican Revolution' in Understanding Information

Ronald E. Day

The Traditional Model of Documentation and the Modern Conception of Information

The traditional understanding of documentation and now the modern conception of information as an 'information and communication model' (Frohmann, 2004a, 2004b) relies upon a metaphysical understanding of being in terms of the ontological presence of entities and their epistemic representation (information), both understood as substances. It originates in Plato's theories of forms and of knowledge as acts of remembrance (*anamnesis*), where knowledge representation (information) reflects the essential nature of beings, like subject headings reflect the essential content of books. Traditional documentation techniques, such as category and classification assignment, and more recent configurations of information through algorithms and now generative AI, lead to the understanding of information as representational assertions about entities and events. These assertions are 'about' entities. In Library and Information Science and documentary Knowledge Organization generally, such 'aboutness,' such 'something about something,' is usually expressed through class properties for entities via categories and descriptors. For example, book y is about dogs, and animal x is (about being) an antelope. Information is assumed to be a true representation of an entity or an event. Such substantive understandings of information extend to folk theories of communication and language, understood as the transfer of these representations—'information'—from a 'sender' to a 'receiver' (the 'conduit metaphor' (Reddy, 1979)).[1]

A different approach and tradition than that of the information-communication model, also in Library and Information Science and the documentary sciences, emphasises context as a condition for meaning, and so, for what is

DOI: 10.4324/9781003583424-5

informational. This tradition of hermeneutics conceives of information as the meaning given to entities and events by cultural forms and the social norms for using them. Cultural forms (such as language) and the social norms by which we use them give us 'worlds' of possible meaning for entities and their relations (Heidegger, 1962), worlds where entities are recognised as such and so are informational about themselves and their relations to other entities. Within an Aristotelian tradition, entities ('beings') are understood in their world in theoretical and pragmatic manners: Theoretically in terms of properties and pragmatically according to how they relate to one another and how we use them (Heidegger, 1962). Whereas, from a metaphysical perspective, entities are a given, and their representations agree (correspond) or not with them (in truth or falsity), hermeneutically conceived entities *are* (they have 'being') due to the cultural and social understandings and use we give to them through language and tradition.

Traditionally, bibliography and documentation arose out of the study of texts, and so the very term 'documents' is still often associated with books, journal articles, and other paper and digital textual forms. With these textual forms, the 'aboutness,' or 'information,' of the entity is suggested by our interpretation of the document. With texts, such interpretations are strongly conditioned by shared hermeneutics—shared cultural traditions and language that we bring to reading and writing texts. If a cataloguer assigns to a text the descriptor—the aboutness—'dogs,' for example, it is likely that the text discusses animals that wag their tails and bark. The successful or 'true' correspondence of the descriptor to the text, of the (information) representation to its object, is likely because of a shared understanding. But not all documentary entities are so agreeable or malleable for representation as texts. (Nor are all texts, of course, and so for this reason some texts, such as fiction, may just be given genre subject representation, or in the case of craftworks, form and function representations).

In her 1951 book, *Qu'est-ce que la documentation?* (Briet, 1951, 2006), the French documentalist Suzanne Briet chose not a text but an animal, what would become a new species of antelope, as her exemplar for a document. This is notable, not least, because it has led to a host of later 'neo-documentalist' commentators, who have highlighted this work and her example (Buckland, 1991, 1997; Day, 2001, 2024; Fremery, 2024; Fremery & Buckland, 2022; Frohmann, 2012; Gorichanaz & Latham, 2016; K. Latham, 2012, 2020; K. F. Latham, 2014). What is so significant in her example that it has caused a new theory of documentation, and so a new philosophy of information, to be built around it?

To answer this question, we turn to an earlier 'neo-documentalist,' Suzanne Briet's student, Robert Pagès, who wrote a dissertation and then a subsequent article published in 1948, entitled 'Transformations documentaires et milieu culturel' (Pagès, 2021b), which Michael Buckland has recently translated as 'Documentary Transformations and Cultural Context' (Pagès, 2021a). In his article, Pagès calls for a new theory of documentation in light of 'new'—that is, non-textual—objects of documentation. Animals and scientific collections are highlighted in his article, but also historical artefacts, such as 'Napoleon's hat.'

In Pagès' article, several important characteristics of these 'new' (or more broadly encompassing) documentary objects, which he characterises as 'auto-documents,' are presented. First, Pagès calls for a new theory or philosophy of documentation—beyond Paul Otlet's grounding of 'documentology' in bibliography and the culture of 'the Book' (Otlet, 1934)—that focuses on documentary entities as particulars, characterised by their 'uniqueness.' Second, such 'uniqueness' refers to not only the newness of entities or kinds of entities to already established classes or collections, but it also refers to the historicity, that is, the historical powers and specificity—expressions and resistance to interpretations—that an entity has by virtue of its own properties (movement and growth in natural entities) and/or its historical provenance (e.g., Napoleon's hat, as either a particular hat that Napoleon owned or as a kind of hat made famous through its being worn by Napoleon). Third, the entity appears as documentarily evident not only through itself but also by virtue of the addition of a 'symbolic set.' All these are characteristics of what he calls the 'auto-document'—the document as a powerful particular—which seems to be, for Pagès, the foundational concept of his new theory or philosophy of documents and documentation. Pagès writes the following, contrasting traditional documentation (as a type-token theory of universal classes applied to particulars) with a theory of the auto-document as a powerful particular. Below, I use Buckland's translation with my own comments in brackets, followed by the French original:

Specimens and Auto-Documents

> **460**
>
> A "unique object" or a member of a collection of similars perhaps kept and used as a "document" about the domain to which it belongs, or even about its own objective properties as (1) an individual or (2) of a type.
>
> In the second case it is matter of a specimen: an anonymous Egyptian mummy, a gorilla in a cage, a piece of rock crystal.
>
> In the first case the document transmits information about itself (Napoleon's hat, a unique meteorite). It is, therefore, an "auto-document." In fact, it is so only through association with [perhaps better translated by, "it—the auto-document—is such only by means of the addition of"] verbal or more elaborate symbols (name, dates, historical description, classification code) which connect it to different symbolic sets.
>
> (Pagès, 2021a)

Échantillons et Auto-Documents

> 460
>
> Un "objet **unique**" ou membre d'une **collection de semblables** peut être conservé et servir de "document" sur un domaine auquel il se rattache, ou bien sur ses propriétés objectives directes, (1) individuelles ou (2) spécifiques.
>
> Dans le second cas il s'agit d'un **échantillon**: une momie anonyme d'Egypte, un gorille dans une cage, un morceau de Spath.
>
> Dans le premier cas le document transmet des renseignements sur lui-même (chapeau de Napoléon, aérolithe singulier). C'est alors un **"auto-document"**. En fait, il n'est tel que moyennant l'adjonction de symboles verbaux ou plus élaborés (dénomination, dates, historique descriptif, cote classificatoire) qui le relient à des ensembles symboliques différents.
>
> (Pagès, 2021b)

Neo-Documentation and Recent Philosophy

There are several interesting intersections of recent theory in neo-documentation and modern philosophy with Pagès' theory. All these involve a critique of representational theories of being and ontology, and thus of information as representation.

First, Michael Buckland's 1991 article 'Information as Thing' proposes that informational artefacts (later in his works, documents), such as Suzanne Briet's antelope, are not just representations of universal class properties but themselves demonstrate 'representativeness' (Buckland, 1991). Such representativeness appears as entities are introduced into collections. Later works, such as his article 'Documentality Beyond Documents' (Buckland, 2014), propose artefacts such as passports and electronic door keys as being not just representational of the bearer but as powerful particulars more powerful than the bearer, insofar as they can empower the bearer by their powers within documentary systems. In all these cases, the document is presented as itself having powers, as organically, technically, or socially given.

Earlier than Buckland's (2014) article, in a 2012 article, 'The Documentality of Mme Briet's Antelope' (Frohmann, 2012), Bernd Frohmann deployed the term 'documentality' as a philosophical term for signifying documents as powerful particulars within contexts that act as affordances for their expression. Frohmann (2012) even proposes that animals themselves are document-producing, in that in their environments they produce traces (urine markings and plumage), independent of any human being's presence.

These are powerful theories in several ways. First, though they were written much later and without knowledge of Pagès' article, they recall Pagès' earlier concept of the 'auto-document.' Second, they propose that the document's documentality—its actual powers of evidentiality—don't lie in the document's belonging to established categories, but instead that the document has, within collections of entities that act as affordances, powers of evidential expression that lie outside of the information-communication model and its metaphysical assumptions upon which much of documentation and information science theory rests, and even for Frohmann (2012) outside of human signification. (Passports, driver's licences, and wedding licences are primarily performative, not representational, documents. Their performances are constitutive of their 'documentality,' to use this term in Ferraris' sense (Ferraris, 2013).) As such, documentation should be discussed outside of the 'information-communication' model of representation (senders and receivers, information as 'epistemic content' in physical or digital forms, etc. (Frohmann, 2004a)). Third, the evidentiality that we associate with documents is not only epistemic but also ontological (such as the animal's environment). As Bruno Latour and Steve Woolgar have shown (Latour, 1986), Louis Pasteur designed experiments where yeast could express itself as a powerful particular. But such expressions required not just the disposition of yeast but also the affordances of laboratory apparatuses, theories of fermentation, and airborne microorganisms. If the setting for documentation—for demonstration—were expanded to the laboratory (the etymological roots of 'documentation,' *docere*, mean 'to show' (Lund, 2024)), then these would all be parts of Pagès' documentary 'sets,' which are assemblages of dispositions for affording the yeast's expression of its being. Fourth, therefore, 'documentality' doesn't lie only in the so-called Knowledge Organisation or Knowledge Representation techniques of libraries or even documentation centres per se, but such techniques belong to larger research cultures (as both Pagès and Briet argued), as well as occurring in natural environments. All this constitutes a radical break from what Pagès (2021a, 2021b) saw as the mediaeval semiotic and ontotheological origins of traditional documentation.

The issue then is where the origin of a document's evidential meaning—for documentality—is found? In Platonic metaphysics, which is the metaphysical origin for traditional documentation's theory of aboutness, the evidential meaning of an entity is its belonging within a universal class because it shares essential properties belonging to that class. However, Pagès, Buckland, and Frohmann's works suggest that a document's documentality—its evidentiality—lies in two elements: Documents as powerful particulars and the assemblages through which such are expressed. That is, documents as dispositions and context or collections as affordances for the evidentiality of powerful particulars.

The relation between a material entity's ontological documentality (its phenomenologically based appearance or evidentiality) and its epistemic documentality (its being evidence through a process) lies in an allegorical or metonymical relation between the two, where the former appears at certain points of congruence, like bringing a pair of binoculars into focus and allowing an entity to clearly appear. It is shown in the manipulation of information systems and the positioning of entities in laboratory experiments through the adjustment of variables or through changes in perspective and methods in field observation. There, and in everyday information systems, as Buckland (2014) has shown in the case of passports, this congruence between knowledge or information systems and things allows for the passage, empowerment, or blockage of agency and expression of the being of entities understood as powers.

In the philosophical literature, we find similar claims. In his investigation of being, the 20th-century philosopher Martin Heidegger located the expression of things as beings in their being gathered or collected with others. Such collections lead to the event [*Ereignis*] of the presence and expression of being. And in the works of the philosopher of science Bruno Latour, socio-technical assemblages are gathering sites or collections for 'articles' (documentarily understood and etymologically referring to artefacts and beings) towards an entity or entities' *artic*ulations (Latour, 1987, 1999, 2005).

So, what is the figurative and literal importance of the animal in all of this? Obviously, the animal as a powerful particular with an expressive force and resistance of its own is an exemplar of Frohmann's (2012) understanding of the document as having force and resistance. Such an entity shows itself (is auto-documentary) in both its own and epistemic environments. Even in *Qu'est-ce que la documentation?* Suzanne Briet mourns that 'the documentary fertility'—the potentiality of evidentiary power—of her newly found animal becomes buried under a veil of documents by traditional documentation once it is captured and confined in a zoo and is subject to a documentary classification as a new type of antelope.

Historically viewed, we can understand Pagès and neo-documentation's turn towards a new philosophy of documentation—and a new philosophy of information—as akin to early 20th-century anthropology and ethology's turns from collections-orientated museology to fieldwork (Debaene, 2014). In all of these, there is an attempt to understand documentality not in terms of a priori categories applied to entities, but instead in terms of contexts acting as affordances for entity powers.

But as Pagès' example of Napoleon's hat suggests, such a turn towards a new theoretical understanding of documents, documentation, and information doesn't only end with a greater emphasis upon natural science documentation and their entities. As the works of the above authors show, this new view, that of neo-documentation, can also be turned back upon cultural and textual documents, with a renewed appreciation of principles such as provenance and fonds in archives and rare books. Digital records have allowed for the broader subject cataloguing of archival documents, but these don't supplant (and sometimes they cover up) the historical specificity and powers of documents that archival arrangement attempts to amplify. And rare books have their 'information' by dint of their documentary form, which is thoroughly historical. Such documentary materials draw in and project historical forces beyond whatever the document's contents were thought to have assumed, even by their creators. Viewing these textual documentary forms and their institutions within the neo-documentalist perspective gives us a new view of them.

The figure and the reality of the animal in documentation force us to critically reconsider documentation and information in terms of the information-communication model and its metaphysical tradition. From the perspective of neo-documentation, an entity's evidentiality is due not to representation but to a mutual appropriation and struggle of powers for expression, to a phenomenology and process of appearance that occurs through ontological and epistemic assemblages or collections. Information systems no longer simply represent entities, but rather entities are now empowered or disempowered through the affordances of such systems. The systems, too, however, as in scientific experiments and other practical uses, need to be adjusted—empowered or disempowered (again, as with binoculars, adjusted in terms of their power of magnification and vision parameters)—so that entities can express themselves and appear. All new information systems dealing with organic agencies or complex systems (such as machine learning) need to be 'parametric' and relational in this way—a documentality beyond traditional document (that is, textual) forms, as Michael Buckland (2014) put it, and beyond traditional documentation and information science based on knowledge representation. This is the new 'cultural necessity' (Briet, 1951, 2006) to which neo-documentation responds and which makes 'the animal' the exemplary documentary figure in its theoretical writings.

Note

1 See also Erik Radio's (2026) critique of the representational model of documentation and information based on the particular (conceived dialectically, as per Adorno) and upon absence in this volume, 'Massive Unseen Companions and the Documentary Project,' Chapter 14.

References

Briet, S. (1951). *Qu'est-ce que la documentation?* Paris: Édit/Éditions Documentaires.

Briet, S. (2006). *What is documentation?: English translation of the classic French text.* Lanham, MD: Scarecrow Press.

Buckland, M. K. (1991). Information as thing. *Journal of the American Society for Information Science*, *42*(5), 351–360.

Buckland, M. K. (1997). What is a document? *Journal of the American Society for Information Science*, *48*(9), 804–809.

Buckland, M. K. (2014). Documentality beyond documents. *The Monist*, *97*(2), 179–186.

Day, R. E. (2001). *The modern invention of information discourse, history, and power.* Carbondale, IL: Southern Illinois University Press.

Day, R. E. (2024). *Foundational writings.* Sacramento, CA: Library Juice Press.

Debaene, V. (2014). *Far Afield: French anthropology between science and literature.* Chicago, IL: University of Chicago Press.

Ferraris, M. (2013). *Documentality: Why it is necessary to leave traces.* New York: Fordham University Press.

Fremery, W. D. (2024). *Cats, carpenters, and accountants: Bibliographical foundations of information science.* Cambridge, MA: MIT Press.

Fremery, W. D., & Buckland, M. K. (2022). Context, relevance, and labor. *Journal of the Association for Information Science and Technology*, *73*, 1268–1278.
Frohmann, B. (2004a). *Deflating information: From science studies to documentation*. Toronto: University of Toronto Press.
Frohmann, B. (2004b). Documentation redux: Prolegomenon to (Another) philosophy of information. *Library Trends*, *52*(3), 373–407.
Frohmann, B. (2012). The documentality of Mme Briet's antelope. In J. Packer & S. B. Crofts Wiley (Eds.), *Communication matters: Materialist approaches to media, mobility and networks* (pp. 173–182). New York: Routledge.
Gorichanaz, T., & Latham, K. (2016). Document phenomenology: A framework for holistic analysis. *Journal of Documentation*, *72*(6), 1114–1133.
Heidegger, M. (1962). *Being and time*. New York: Harper & Row.
Latham, K. (2012). Museum object as document: Using Buckland's information concepts to understand museum experiences. *Journal of Documentation*, *68*(1), 45–71.
Latham, K. (2020). Two-headed calves, unicorn horns, and trephined skulls: An essay about beautiful museum monsters. *Proceedings from the Document Academy*, *7*(1). Retrieved from https://ideaexchange.uakron.edu/docam/vol7/iss1/2/
Latham, K. F. (2014). *Foundations of museum studies: Evolving systems of knowledge*. Santa Barbara: Libraries Unlimited, an imprint of ABC-CLIO, LLC.
Latour, B. (1986). *Laboratory life: The construction of scientific facts*. Princeton, NJ: Princeton University Press.
Latour, B. (1987). *Science in action: How to follow scientists and engineers through society*. Cambridge, MA: Harvard University Press.
Latour, B. (1999). *Pandora's hope: Essays on the reality of science studies*. Cambridge, MA: Harvard University Press.
Latour, B. (2005). *Reassembling the social: An introduction to actor-network theory*. Oxford: Oxford University Press.
Lund, N. W. (2024). *Introduction to document studies*. London: Facet Publishing.
Otlet, P. (1934). *Traité de documentation: le livre sur le livre: théorie et pratique*. Brussels: Editiones Mundaneum, Palais Mondial.
Pagès, R. (2021a). Documentary transformations and cultural context. *Proceedings from the Document Academy*, *8*(1). https://doi.org/10.35492/docam/8/1/2
Pagès, R. (2021b). Transformations documentaires et milieu culturel (1948). *Proceedings from the Document Academy*, *8*(1). https://doi.org/10.35492/docam/8/1/3
Radio, E. (2026, this volume). Massive unseen companions and the documentary project. In N. Solhjoo (Eds.), *Multispecies information science*. Routledge.
Reddy, M. J. (1979). The conduit metaphor: A case of frame conflict in our language about language. In A. Ortony (Ed.), *Metaphor and Thought* (pp. 284–310). Cambridge: Cambridge University Press.

5

WHEN ANIMALS TEACH

Lessons for Information Pedagogy

Niloofar Solhjoo and Jenna Hartel

Introducing Multispecies Pedagogy

Incorporating animals into information pedagogy is about transformation. It challenges the ontological and epistemological foundations of the discipline and invites more relational and responsible modes of knowledge production (Lloro-Bidart, 2018). Multispecies pedagogy refers to educational approaches that integrate the agency, voices, and relationships of humans and nonhuman beings (e.g., animals, plants, and ecosystems) in the learning process (Ameli, 2024; Nivala & Ryynänen, 2025).

Multispecies pedagogy has been developed across various disciplines such as education, early childhood learning, and human–animal studies. There are three key characteristics that define this approach. First, learning is understood as co-created in collaboration with nonhuman actors, such as classroom dogs or urban wildlife. Second, nonhumans and nature are treated as active participants in educational settings, rather than as objects, symbols, or passive tools (Kozak, 2024; Ogden et al., 2013). Third, multispecies pedagogy promotes a holistic approach that fosters empathy and justice. For example, in an animal-assisted intervention in a library, a dog is not merely a tool for children learning but a co-educator whose stress signals—such as panting or yawning—must be carefully observed, learnt, and respected.

Learning with animals is a holistic process that involves moving through experience, reflection, conceptualisation, and action—allowing students to form deeper, embodied connections with course content (Hovorka, 2019). As Ringrose and Renold (2016) assert, nonhuman animals can become co-creators of knowledge in educational settings.

DOI: 10.4324/9781003583424-6

When students encounter animals—whether through stories, photographs, guest speakers, or their own companion animals—the classroom becomes a space of entangled agency, where learning is collaboratively shaped. These pedagogical practices echo multispecies ethnographies that highlight how human–animal relations produce knowledge that is affective and situated (Pacini-Ketchabaw, Taylor & Blaise, 2016).

To support engagement with animals in curricula, educators across different disciplines employ a range of interdisciplinary, experimental, and creative methods. These approaches span from subjective and qualitative to more scientific and measurable ones. In Chapter 6, Solhjoo (2026) outlines multispecies methodologies that can inform the development of multispecies pedagogy and teaching initiatives.

Learning to Welcome Animals in the Library

As animals increasingly come into public spaces, including libraries, new content is needed in the Information Science curriculum. The idea can be planted and developed that animals may be guests of a library or our helpers in delivering programmes; either way, best practices are necessary. Here, we survey five such examples and point to resources for instructors and learners alike.

Animal Policies in Libraries

Future librarians should know what qualifies as a service animal, therapy animal, and emotional support animal, and what these categories mean for libraries. To explain, *service animals*, an official designation which includes only dogs and miniature horses that assist people with disabilities, are allowed into public spaces per legislation in most regions of the world. Differently, the latter categories of *therapy* and *emotional support animals* (which may include a wide variety of animals) are not legally mandated for entry. These animal roles are further subject to local policy determined within a library and must be communicated consistently through signage and the enforcement of library staff. Patrons with dependencies and great affection for their therapy or emotional support animals may insist upon their admission to a public building. Librarians must learn during their training to respond in a compassionate, professional, and management capacity. The succinct article, 'The Importance of Establishing Assistance Animal Policies for Your Library' (Rust & Wise, 2017), is an excellent introduction to these issues.

Reading Assistance Animals

An information literacy programme, increasingly popular in public libraries and schools, involves children reading to dogs. The concept of *reading education assistance dogs* was hatched by Intermountain Therapy Animals in 1999. The practice spread as an offering from certified therapy dog organisations, which provide the dog and a trained handler. At the session, a therapy dog is available in a quiet space of a library or school. Children sign up for an appointment and may read out loud to the dog, who listens non-judgementally (and may cuddle, too). Advocates argue that reading assistance education dogs help children learn to read by increasing relaxation, confidence, and motivation. 'Children Reading to Dogs: A Systematic Review of the Literature' is an expert introduction to this topic (Hall et al., 2016). The paper contains a model of the process and confirms that positive impacts are plausible, though more careful scientific measurement of the practice is needed. Such programmes as these would fit into courses or lessons on information literacy instruction, children and youth services in libraries, or school libraries.

Therapy Animal Programmes, Real, and Robotic

Another animal-assisted intervention is increasingly occurring at college and university libraries. During stressful periods, such as examination times, students are invited to interact with a therapy animal as a respite from work, to reduce anxiety, and to increase well-being and sociality. While seasoned therapy dogs typically star in such programmes, cats, rabbits, and even mini-horses may be encountered at the library, too. The positive benefits of interacting with animals have been well-documented (Levinson, 1964), and there are contemporary statements that declare a new paradigm, too (Fine et al., 2019). 'Studying for Exams Just Got More Relaxing—Animal-Assisted Activities at the University of Connecticut Library' is a great case study that outlines the rationale, communication, and logistics for success (Reynolds & Rabschutz, 2011). Courses in academic librarianship and information literacy would benefit from a primer on such programmes.

Beyond academic libraries, therapeutic animal helpers can take surprising forms. Public libraries in New South Wales, Australia, have introduced MetaCats into educational and communal spaces. Librarians work with these robotic cats to provide therapeutic outreach services, particularly for children and older patrons living with dementia (The Guardian, 2025). Equipped with an artificial heartbeat and non-allergenic fur, the robotic cats simulate lifelike cat behaviours and play a meaningful role in inclusive library programming.

Storytelling with Animals and More

Animal-themed collections and programmes for children are common in libraries. This is because animals are important and effective educators for our young people. Melson's (2001) pioneering book, *Why the Wild Things Are: Animals in the Lives of Children*, comprehensively outlines the pervasiveness of animals in children's worlds; the deep emotional attachments children form with pets; how interactions with animals stimulate young people's cognitive growth as well as social and moral development; the therapeutic power of contact between children and animals; and the powerful symbology of animals in children's imaginations, dreams, and play. 'The Animal as Fourth Educator: A Literature Review of Animals and Young Children in Pedagogical Relationships' (Bone, 2013) is a gateway to learn more.

Storytime, a long-standing offering of public libraries, is rich with animal characters and themes. Indeed, the most popular children's book ever, 'The Very Hungry Caterpillar,' stars a voracious insect. The effectiveness of animal motifs transcends entertainment. Early literacy often begins with learning animal names, and their unique properties are a fantastic way to teach vocabulary: A giraffe is *tall*, a cheetah is *fast*, and an octopus has *eight* arms. The sounds animals make—from the *roaaarrr!* of a lion to the *moooooo-oo!* of a cow—are endlessly fascinating to children and immediately generate laughter and hilarity. We can imagine a lesson in a Library and Information Science (LIS) programme where students practise animal sounds together to gain confidence as creative, uninhibited, multispecies storytellers.

Beyond storytime, animals spark many other types of programming. Some libraries host chick-hatching events, where children can witness the emergence of a new brood and celebrate their arrival in the world. Another trend at public libraries is a stuffed animal sleepover. Kids drop off their soft companions for a night of stories and fun, picking them up the next day along with photos and videos that double as heartwarming promotional material. One of the authors of this paper teaches a six-week workshop, 'Twisting Balloon Animals,' where future librarians master a menagerie of balloon forms which serve as gateways to animal information, conversations, and consciousness. After adding reader's advisory and puppet shows to this round-up, it is easy to imagine an entire course on animal-themed adventures for children and families in libraries.

Greening the Library: Earth-Aware Programmes and Advocacy

Today, with the planet imperilled, libraries are taking a stand for a wide range of environmental issues. In 2015, the American Library Association (ALA) adopted sustainability as a core professional value, a commitment it reaffirmed in 2019. This dedication was deepened in 2023 with the launch of the ALA National Climate Action Strategy. In practice, libraries can aim to be green buildings that function as models for their communities.

As neutral and safe spaces, libraries can also host lectures, exhibitions, and reading groups that foster sober reflection on contested public issues such as climate change. Beyond these important discussions, information institutions can offer hands-on, all-ages programmes that are both educational and fun. For example, workshops on composting teach patrons to transform kitchen scraps into rich soil; gardening tutorials help families to grow their own food; 'upcycling' events inspire people to turn everyday trash into new treasures; and a 'Repair Café' aims to extend the lifespan of sundry devices that would otherwise be repurchased. The skills needed to plan and execute such sessions are as vital as those for information literacy instruction or reference services. By opening the eyes of students to a range of Earth-aware topics, LIS educators can prepare future librarians to be catalysts for environmental action that sustains our Earth as a home for all species.

Imagining the Beyond-Human LIS Curriculum

LIS engages with questions of inclusivity, ethics, and epistemic justice, particularly in relation to systemic inequalities and the need for critical approaches to diversity in education (Hudson, 2017). A multispecies perspective offers a compelling framework for reimagining both curriculum and teaching practices. Here, we elaborate on where and how a multispecies perspective is presented and integrated into the LIS curriculum.

Foundational Courses

A multispecies approach can function as a metatheory within foundational courses, complementing the 13 proposed by Bates (2005). The unique contribution of a multispecies perspective in our curricula is to challenge dominant human-centred narratives by incorporating animal and environmental realities and to supply a new way to conceptualise knowledge, relationality, and learning beyond the human (Lloro-Bidart, 2018; Nivala & Ryynänen, 2025).

Information Science is arguably difficult to teach given its unique concern for information phenomena that are embedded within the surface variety of life. Bates (1999) argues that it takes neophytes a few months of training, and sometimes much longer, to grasp our disciplinary outlook and foci. The familiar and congenial space of the animal kingdom can serve as a resource to demystify some of our more intangible and intimidating concepts.

For starters, let us simply imagine metaphorical crossroads between the animal kingdom and Information Science. The 'metaperspective' (conceptual distance) needed to see information structures and patterns may be likened to the vision of high-flying birds, who soar over a landscape and with 'eagle eyes' can discern the movements and contours of things far below. Birds also illustrate information migration and network patterns. Antelopes demonstrate adaptive information behaviour in response to environmental pressures and represent a role model in animal turn in traditional LIS (see Chapter 1). The practice of using both print and digital reference sources, which involves skilful transverse across the fields within a record (while leaping over much of the content), can be likened to the spirited 'pronking' of gazelles, who spring swiftly and knowingly over and across the land. The echolocation abilities of the bat resemble the process of an information professional navigating an online information collection—both beings are highly successful at finding something hidden in the dark. These metaphors help materialise abstract LIS concepts through embodied, multispecies relations.

Beyond metaphor, the universal view of information processing (Bates, 2005) supports an expanded pedagogy for learning the foundations of information studies. Even simple organisms display remarkable information behaviours: Insects communicate ecological risks (Anton et al., 2013), tree root networks share environmental information like distributed brains (Ciszak & Mancuso, 2013), and microbes use chemical cues and collective memory for decision-making (Clark, 2013). This multispecies perspective emphasises that information, knowledge, and learning are not exclusive to humans but are distributed across more-than-human worlds, supporting a richer, ecologically grounded LIS pedagogy.

Critical Librarianship and Contemporary Issues

There is a clear need to strengthen the philosophical foundation of LIS by integrating critical theory into its curricula. While librarians are actively engaged in social justice work—improving access, supporting marginalised communities, and promoting inclusive practices—their theoretical training often occurs outside of LIS programmes (Schroeder & Hollister, 2014). Integrating a multispecies perspective offers a uniquely powerful way to address this gap, as it is a critical theory that is both intellectually rigorous and broadly inclusive.

Unlike some critical traditions that can be perceived as obscure or elitist (Beilin, 2018), a multispecies perspective is grounded in a universal experience: Humans have always lived and learnt alongside animals. One does not need a PhD to understand the loyalty of a dog or the curiosity of a cat. This familiarity provides a low barrier to entry to critical librarianship, making complex ideas about power, care, and relationality intuitively accessible. It can unite people across academic and non-academic divides, using shared concern for the world as a less divisive starting point for conversations about justice and ethics.

Multispecies pedagogy resonates with critical librarianship and contemporary issues. Practical programmes, such as those discussed in the *Learning to welcome animals in the libraries* section, are perfect examples: They require no theoretical jargon yet demonstrate care and inclusivity in a tangible way, embodying the profession's core values of equity and access.

By valuing non-verbal, empathetic, and embodied ways of knowing (discussed in the *Foundational courses* section), the multispecies perspective honours forms of intelligence often neglected by academic discourse. Thus, it serves as an accessible tool for liberation and solidarity, empowering librarians and LIS students to challenge systemic inequalities, transform their practice, and ultimately advance a more inclusive and critically reflective profession.

The multispecies turn directly addresses the tension identified by Beilin (2018) by offering a form of theory that is intellectually challenging yet immediately relatable, helping to break down the false dichotomy between theory and practice that often excludes marginalised voices. As critical theories associated with race and Indigenous experience rightly gain visibility in LIS for both research and pedagogy (Pawley, 2006; Roy, 2015), some readers may question whether animals merit the field's limited activist attention and resources, especially when so many humans continue to face systemic injustice. However, we do not see these concerns as mutually exclusive. Rather than competing, these struggles can and should reinforce one another as part of a broader, integrated effort to confront intersecting systems of oppression.

Indigenous Ways of Knowing

Multispecies pedagogy brings approaches that respect other ways of knowing, where animals are understood as teachers and kin. As Lloro-Bidart (2018) argues, interrogating the human/animal divide can help deconstruct systems of marginalisation that have historically excluded both animals and certain human communities from knowledge production. In this sense, the multispecies perspective naturally honours Indigenous knowledge systems, which understand the world through reciprocity and relationality rather than extraction and hierarchy (Nxumalo & Villanueva, 2019; Todd, 2016). This alignment between multispecies and Indigenous perspectives has already provided a philosophical foundation in fields such as justice. Building on Indigenous knowledge and lifeways, the concept of *multispecies justice* advances an inclusive framework that affirms all living beings as deserving of respect and protection, while holding humans accountable for the harms they cause to the natural world (Winter, 2022). For instance, Māori philosophy does not recognise a strict living/non-living dichotomy but views animals as guides and even signatories to treaties (Jones & Hoskins, 2016).

Knowing, in this context, is a process of being in respectful relation, not one of mastery. This theoretical alignment can be applied in LIS education. In our curriculum, we integrated an activity centred on the *Pepeha*, a Māori introduction that situates one's identity in relation to their mountain, river, tribe, and ancestors. This activity not only introduces non-Indigenous students to Indigenous ways of knowing but also encourages them to reflect on their own experiences through a relational lens connected to the land and the more-than-human community. For the Indigenous students, the Pepeha resonated differently—it became a moment to reflect on the disruption caused by colonisation, including the loss of family names and displacement from their lands.

Practical Teaching Resources

For educators and students interested in incorporating multispecies perspectives into LIS education, several resources stand out.

i **Readings**: Two chapters in this volume are particularly helpful. *Chapter 1: The Footprints of Animals in Library and Information Science* by Solhjoo and Hartel (2026) traces the evolving role of animals and nonhuman organisms in LIS, exploring how they inform and change traditional LIS and documentation, develop information studies and frameworks, and knowledge organisation. *Chapter 7: Methodological Pathways for Multispecies Information Science* by Solhjoo (2026) introduces both practical approaches to learning, working and living with animals and more-than-human assemblages within information disciplines, offering practical guidance for integrating multispecies into teaching and research design.

ii **Multimedia**: Free videos can complement readings and appeal to visual learners. At INFIDEOS (Hartel, n.d.), the 3.5-minute video *Multispecies Perspective* introduces its subject at a very high level. The video surveys key terms (e.g., anthropocentrism, human exceptionalism, and entanglements); recognises the absence of animals within Information Science to date; and showcases the work of information scientists who champion the perspective, namely Marcia J. Bates and Niloofar Solhjoo. The short and lively video is right-sized for watching in class and may serve as a multimedia gateway to writings that are now emerging on the topic. For more in-depth content, see the two-part lecture series *SICS Seminar: Introduction to Multispecies Information Science*, delivered by Niloofar Solhjoo at Charles Sturt University. It is available on the YouTube channel of the School of Information and Communication Studies (SICS) and provides extended multimedia material suitable for integration into LIS curricula (Figure 5.1).

Fostering Multispecies Connection in the Classroom

Small Approaches

Educators apprehensive about or new to multispecies pedagogy might simply take advantage of the universal affection students share for animals and nature to spark friendship, positivity, and a multispecies moment in class. To that end, below are three playful, tested activities that build classroom rapport while

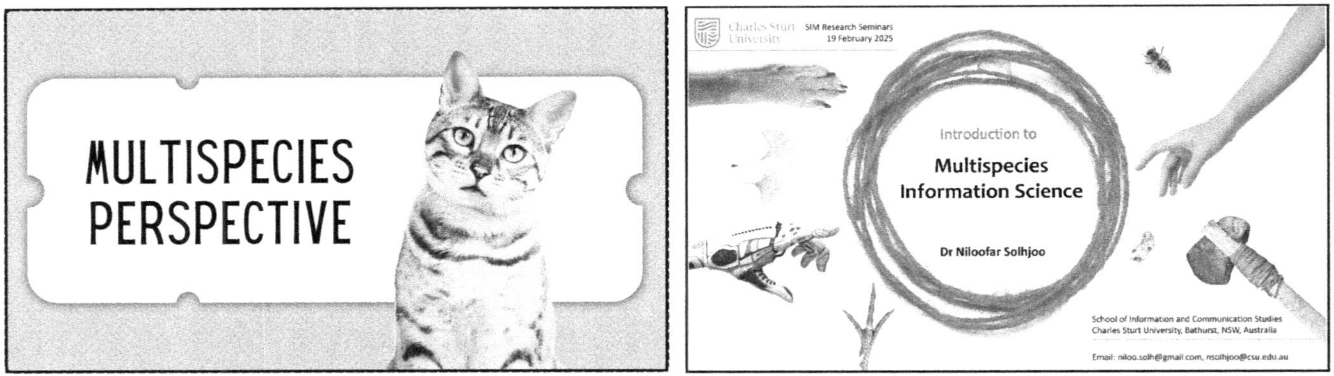

FIGURE 5.1 The thumbnail from the videos, *Multispecies Perspective* (Hartel, 2023) and *Introduction to Multispecies Information Science* (Solhjoo, 2025a, 2025b).

gently laying the foundation for multispecies thinking. These are great starting points for instructors who have never invited animals into their classrooms.

i An icebreaker at the first gathering of the semester asks everyone to think of their favourite animal and three of its redeeming qualities, which are shared with peers in small groups (or altogether). The instructor then reveals a deeper symbology—these preferences reflect each individual's most precious *human* qualities, leading to a heartwarming moment of appreciation among classmates. This activity uses attractive and signal features of animals to help students recognise positive self-conceptions, which they might otherwise be unwilling to voice; it has roots in Freud's psychodynamic theory, as well as symbolism and archetype theories from Jungian psychology.

ii To illustrate: The first author of this chapter, Niloofar Solhjoo, her favourite animal is a dog, because of their care, cheerfulness, and the way they like to share everyday moments and activities with humans. Friends and colleagues can attest that Niloo is often this way. The second author of the chapter, Jenna Hartel, her cherished animal is a dove, on account of their gentleness, stillness, and soaring ability—compliments that are sometimes granted by students in her teaching evaluations.

iii During a pause within a lecture or a break between class segments, students of an online synchronous class can be invited to bring pets into view on camera and to provide introductions if there is time. Instantly, smiles and laughter erupt, and everyone is reminded of the joy animals bring into our lives. What is more, we see the depth and variety of life existing outside the classroom frame, and the seeds for many subsequent conversations among classmates are planted.

iv In the tradition of contemplative pedagogy (Barbezat & Bush, 2014), short restorative meditation can occur in an online or in-person class (Hartel et al., 2017). One particular approach is a progressive loving-kindness meditation (*metta bhavana*). The process begins by imagining and engendering loving feelings towards the self and then extends outwards to all beings in the universe (an inherently multispecies perspective). A category that receives special loving attention along the way is any adorable animal, such as a puppy or kitten, which meditators are prompted to conjure in the mind. This meditation target is effective because of the unfettered affection that arises from the contemplation of such sweet creatures.

Pedagogical Cases

Next, to demonstrate how a multispecies approach can reshape content and approach in teaching Information Science, here are five pedagogical cases developed and implemented in our university-level information courses. Half of the empirical examples are drawn directly from Solhjoo's (2024) doctoral research with a multispecies family, which used a multimodal, interactive ethnographic methodology—such as photovoice—to examine the informational relationships between companion animals and their humans. The other half derives from the authors' teaching experiences in iSchools.

i In the *Foundations of Information Studies* course, students engage in an activity that involves mapping **information within photographs of animals' daily lives**. The activity encourages students to articulate the implicit, often unnoticed, forms of information that structure daily animal interactions and relationships, mostly with humans. Students explore different forms of the presence of information in everyday life using a curated collection of images—capturing moments from shared lives of cats and dogs with humans. Students begin by selecting a photograph collection—either printed in the classroom or available digitally via *Padlet*—and are guided to map the informational forms embedded in images (Figure 5.2).

ii Creating **animal-related information services** is introduced in the course *Information Resources and Services* using the same visual collections. This exercise invites students to imagine new professional roles that respond to pet care practices and human–companion animal relationships—for example, a 'dog care literacy librarian' or a 'cat art collection curator.' Photographs depicting the process of daily multispecies living help students understand the information produced by all species, which then constitutes the universe of documentation (Bates, 2015), the domain of their work as information professions. These are then categorised according to various aspects of companion animal welfare, drawing on Kari and Hartel's framework of 'higher and lower things' in everyday information studies, as mapped by Solhjoo et al. (2022). Students reflect on information channels and services and information technologies across diverse settings—from mundane, survival-orientated aspects of daily animal care to animals' rich emotional lives, leisure, and enrichment (Figure 5.3).

iii In the course *Information Professions*, the students had the opportunity to engage in **virtual study visits with animal health information practitioners**. Discussions were framed within the history of collaboration between LIS and the veterinary and agricultural studies (Rey et al., 2013) and were further established through the professional association that leads the *International Conference of Animal Health Information Specialists* (Rey et al., 2015). Guest speakers were veterinary academic librarians and veterinary informatics analysts. Working within universities, laboratories, and public libraries, they highlight how their efforts strengthen relationships between practitioners, communities, and animals.

iv The course *User Experience Design* features a project-based activity titled **the national pet passport**. This project-based activity draws students into a speculative yet plausible design scenario of a national dog passport system (Tate and Solhjoo, 2024). Inspired by real-world gaps in animal data infrastructure and informed by fictional but urgent concerns (e.g., zoonotic disease, water toxicity from climate change), the case introduces students to the messy, multifaceted

FIGURE 5.2 The collection (online and physical) of photo diaries documenting companion animals' daily lives is used for mapping information forms and activities.

FIGURE 5.3 Various domains of companion animal welfare embedded in Kari and Hartel's (2007) perspective of higher and lower things in everyday life information behaviour (Solhjoo et al., 2022).

world of pet governance and digital infrastructure. Students apply user experience design thinking tools to create stakeholder analysis and persona development (e.g., dog, dog guardians, veterinarians, local councils) and define the system features such as vaccination tracking, dog's behavioural records, and environmental alerts. Many students imagined an interspecies information system, where data flows from one species (e.g., via wearable devices on dogs) and is processed for interpretation by another (e.g., guardians, veterinarians).

v In any class activity or assignment, space should be made for students to move in multispecies directions, if they wish. For example, participants in a course about information behaviour (IB) are taught to conduct original IB research using the information horizon interview (Hartel et al., 2018). They are allowed to select a population, topic, and IB question of personal interest. Several students of the class, inspired by Solhjoo (2024) and their own passion for animals, pursued information behaviour, practice, and experience that is decidedly multispecies. Accounts of information in bird-watching, dog-walking, the decision to adopt a large dog, immigrating with a pet, managing cat allergies, and navigation while horseback riding were fascinating studies. A sample of information horizon maps from these projects appears below; each illustrates the relationship between people, information, and animals.

Conclusion

Integrating the multispecies pedagogy into LIS is more than adding new content and frameworks—it is a radical call to reimagine the very nature of learning itself. It changes how, what, and with whom we learn. Our pedagogical experiments—from loving-kindness meditations to rethinking curriculum design—demonstrate that this is not an abstract or elitist theoretical exercise. Rather, it is an empathic practice and a tool for solidarity, rooted in a deep respect for nature, animals, and diverse communities. It opens rich opportunities for relational ways of learning, embodied communication, and other forms of intelligence and lifeways. Most importantly, it instils in information professionals a crucial ethical imperative: That their responsibility extends to all beings, human, and nonhuman alike.

When we learn with animals, we learn to care. We learn to connect. Our creativity is animated. It is our hope that the insights and practices presented here will inspire educators across iSchools to embrace the transformative power of multispecies perspectives, integrating them meaningfully into their pedagogy and teaching practices.

References

Ameli, K. (2024). Multispecies ethnography in educational research: Changing perspective in animal-assisted education. In J. Budde, A. Wischmann, G. Rißler, & M. Meier-Sternberg (Eds.), *Novelty, innovation and transformation in educational ethnographic research: European perspectives* (pp. 57–69). Routledge. https://doi.org/10.4324/9781032629384-7

Anton, S., Minoli, S., Gadenne, C., & Anderson, P. (2013). Modulation of olfactory and gustatory sensitivity mediated by sensory experience in moths: A simple form of social learning? In K. B. Clark (Ed.), *Social learning theory: Phylogenetic considerations across animal, plant, and microbial taxa* (pp. 147–162). Nova Science Publishers.

Barbezat, D. P., & Bush, M. (2014). *Contemplative practices in higher education: Powerful methods to transform teaching and learning*. Jossey-Bass.

Bates, M. J. (1999). The invisible substrate of information science. *Journal of the American Society for Information Science*, *50*(12), 1043–1050. https://doi.org/10.1002/(SICI)1097-4571(1999)50:12%3C1043::AID-ASI1%3E3.0.CO;2-X

Bates, M. J. (2005). Information and knowledge: An evolutionary framework for information science. *Information Research*, *10*(4), n4. http://InformationR.net/ir/10-4/paper239.html

Bates, M. J. (2015). The information professions: knowledge, memory, heritage. *Information Research*, *20*(1), paper 655. Retrieved from http://InformationR.net/ir/20-1/paper655.html (Archived by WebCite® at http://www.webcitation.org/6WyNray1L)

Beilin, I. G. (2018). Critical librarianship as an academic pursuit. In K. P. Nicholson, & M. Seale (Eds.), *The politics of theory and the practice of critical librarianship* (pp. 198–210). Litwin Books.

Bone, J. (2013). The animal as fourth educator: A literature review of animals and young children in pedagogical relationships. *Australasian Journal of Early Childhood*, *38*(2), 57–64. https://doi.org/10.1177/183693911303800208

Ciszak, M., & Mancuso, S. (2013). Plant root networks and swarm rules. In K. B. Clark (Ed.), *Social learning theory: Phylogenetic considerations across animal, plant, and microbial taxa* (pp. 163–172). Nova Science Publishers.

Clark, K. B. (2013). The mating judgements of microbes. In K. B. Clark (Ed.), *Social learning theory: Phylogenetic considerations across animal, plant, and microbial taxa* (pp. 173–200). Nova Science Publishers.
Fine, A. H., Beck, A. M., & Ng, Z. (2019). The state of animal-assisted interventions: Addressing the contemporary issues that will shape the future. *International Journal of Environmental Research and Public Health*, *16*(20), 3997. https://doi.org/10.3390/ijerph16203997
Hall, S. S., Gee, N. R., & Mills, D. S. (2016). Children reading to dogs: A systematic review of the literature. *PLoS ONE*, *11*(2). https://doi.org/10.1371/journal.pone.0149759
Hartel, J. [@INFIDEOS]. (2023, December 11). Multispecies perspective (Episode #4 of An Archipelago of Information Science) [Video]. *YouTube*.
Hartel, J. (n.d.). INFIDEOS [YouTube channel]. *YouTube*. Retrieved June 30, 2025, from https://www.youtube.com/@INFIDEOS
Hartel, J., Nguyen, A. T., & Guzik, E. (2017). Mindfulness meditation in the classroom. *Journal of Education for Library and Information Science*, *58*(2), 112–115. https://doi.org/10.3138/jelis.58.2.112
Hartel, J., Oh, C., & Nguyen, A. T. (2018). Teaching information behavior with the information horizons interview. *Journal of Education for Library and Information Science*, *59*(3), 67–79. https://doi.org/10.3138/jelis.59.3.2018-0017.07
Hovorka, A. J. (2019). Teaching the nonhuman animal in higher education: Interdisciplinary experiential learning. *Society & Animals*, *29*(5–6), 559–576. https://doi.org/10.1177/105382590803100104
Hudson, D. J. (2017). On "diversity" as anti-racism in library and information studies: A critique. *Journal of Critical Library and Information Studies*, *1*(1). https://doi.org/10.24242/jclis.v1i1.6
Jones, A., & Hoskins, T. K. (2016). A mark on paper: The matter of Indigenous-settler history. In C. A. Taylor & C. Hughes (Eds.), *Posthuman research practices in education* (pp. 75–92). Palgrave Macmillan. https://doi.org/10.1057/9781137453082_5
Kari, J., & Hartel, J. (2007). Information and higher things in life: Addressing the pleasurable and the profound in information science. *Journal of the American Society for Information Science and Technology*, *58*(8), 1131–1147. https://doi.org/10.1002/asi.20585
Kozak, L. (2024). Fostering an inclusive multispecies social pedagogy: rethinking ethical human–animal relations in art. *International Journal of Social Pedagogy*, *13*(1), 13. https://doi.org/10.14324/111.444.ijsp.2024.v13.x.013
Levinson, B. M. (1964). Pets: A special technique in child psychotherapy. *Mental Hygiene*, *48*(2), 243–248.
Lloro-Bidart, T. (2018). A feminist posthumanist multispecies ethnography for educational studies. *Educational Studies*, *54*(3), 253–270. https://doi.org/10.1080/00131946.2017.1413370
Melson, G. F. (2001). *Why the wild things are: Animals in the lives of children*. Harvard University Press.
Nivala, E., & Ryynänen, S. (2025). An eco-social turn in social pedagogy. *International Journal of Social Pedagogy*. https://doi.org/10.14324/111.444.ijsp.2025.v14.x.006
Nxumalo, F., & Villanueva, M. (2019). Decolonial water stories: Affective pedagogies with young children. *International Journal of Early Childhood Environmental Education*, *7*(1), 40–56.
Ogden, L. A., Hall, B., & Tanita, K. (2013) Animals, plants, people, and things: A review of multispecies ethnography. *Environment and Society: Advances in Research*, *4*(1), 5–24. https://doi.org/10.3167/ares.2013.040102
Pacini-Ketchabaw, V., Taylor, A., & Blaise, M. (2016). Decentring the human in multispecies ethnographies. In C. A. Taylor & C. Hughes (Eds.), *Posthuman research practices in education* (pp. 149–167). Palgrave Macmillan. https://doi.org/10.1057/9781137453082_9

Pawley, C. (2006). Unequal legacies: Race and multiculturalism in the LIS curriculum. *The Library Quarterly*, *76*(2), 149–168. https://doi.org/10.1086/506955

Rey, L. M., Croft, V. F., Whitaker, S. K., & Stephens, G. (2015). Impacting librarianship and veterinary medicine: History of the veterinary medical libraries section of the Medical Library Association from 1974 to 2014. *Journal of Agricultural & Food Information*, *16*(3), 253–270.

Rey, L. M., Croft, V. F., Williams, M., & Boyd, C. T. (2013). A history of the International Conference of Animal Health Information Specialists (ICAHIS) and its partnership in one health. *Journal of Agricultural & Food Information*, *14*(2), 151–163.

Reynolds, J. A., & Rabschutz, L. (2011). Studying for exams just got more relaxing—Animal-assisted activities at the University of Connecticut Library. *College & Undergraduate Libraries*, *18*(4), 359–367. https://doi.org/10.1080/10691316.2011.624934

Ringrose, J., & Renold, E. (2016). Cows, cabins and tweets: Posthuman intra-active affect and feminist fire in secondary school. In C. A. Taylor & C. Hughes (Eds.), *Posthuman research practices in education* (pp. 220–241). Palgrave Macmillan. https://doi.org/10.1057/9781137453082_13

Roy, L. (2015). Advancing an indigenous ecology within LIS education. *Library Trends*, *64*(2), 384–414. https://doi.org/10.1353/lib.2015.0045

Rust, M., & Wise, M. (2017). The importance of establishing assistance animal policies for your library. *PNLA Quarterly*, *81*(4), 16–24.

Schroeder, R., & Hollister, C. V. (2014). Librarians' views on critical theories and critical practices. *Behavioral & Social Sciences Librarian*, *33*(2), 91–119. https://doi.org/10.1080/01639269.2014.912104

Solhjoo, N. (2024). Information experience within multispecies families: Understanding shared life with companion cats and dogs (doctoral dissertation). Open Access Te Herenga Waka–Victoria University of Wellington. https://openaccess.wgtn.ac.nz/articles/journal_contribution/The_In-between_Information_Experience_within_Human-Companion_Animal_Living/24986298

Solhjoo, N. (2025a, February 24). SICS Seminar: Introduction to multispecies information science - Part 1 [Video]. *YouTube*. https://www.youtube.com/watch?v=LRu20pC5RO4

Solhjoo, N. (2025b, February 24). SICS seminar: Introduction to multispecies information science - Part 2 [Video]. *YouTube*. https://www.youtube.com/watch?v=lmDBNHpBdv0&t=433s

Solhjoo, N. (2026). Research pathways for multispecies information science. In N. Solhjoo (Eds.), *Multispecies information science*. (pp. 91–110), Routledge.

Solhjoo, N., & Hartel, J. (2026). The footprints of animals in library and information science. In N. Solhjoo (Eds.), *Multispecies information science*. (pp. 5–21), Routledge.

Solhjoo N, Krtalić M, & Goulding A. (2022). Pets and people: Information experience of multispecies families. *Journal of Documentation*, *78*(5), 1092–1108. https://doi.org/10.1108/JD-03-2021-0052

Tate, M., & Solhjoo, N. (2024). *A national pet passport system for New Zealand* [Case study for course assessment]. Victoria University of Wellington.

The Guardian. (2025, May 28). *Therapeutic robot cats 'a stroke of genius' at Blue Mountains library, Australia*. https://www.theguardian.com/technology/2025/may/28/therapeutic-robot-cats-blue-mountains-library-australia

Todd, Z. (2016). An indigenous feminist's take on the ontological turn:'Ontology' is just another word for colonialism. *Journal of Historical Sociology*, *29*(1), 4–22. https://doi.org/10.1111/johs.12124

Winter, C. J. (2022). Introduction: What's the value of multispecies justice? *Environmental Politics*, *31*(2), 251–257. https://doi.org/10.1080/09644016.2022.2039001

6

THE ETHICS OF MULTISPECIES INFORMATION RESEARCH

An Exploratory Dialogue

Steve Fuller and Finn Lees

Introduction

What follows is an exploratory dialogue on the ethics of multispecies information research between two people who come to the topic somewhat differently. Steve Fuller has been concerned with the normative force of being 'human' given the increasing challenges posed by 'post' and 'trans' humanists, who in their different ways have raised the normative salience of non-human beings. Finn Lees has been concerned more specifically with animal welfare in today's world, where human exploitative tendencies have been extended into relationships with animals. Thus, the dialogue focuses on the extent to which differences between humans and animals should matter when designing an ethics of multispecies information research—and the implications of those differences for the character of such ethics. As will become clear in the two rounds of their exchange, Fuller tends to minimise the difference between humans and animals by leaning on anthropomorphism as the default stance for dealing with animals, while Lees tends to maximise the difference, raising the bar for ethically adequate information transactions between humans and animals.

Although the dialogue ranges widely across several disciplinary perspectives on human–animal relations, it is anchored in the concept of 'auto-documentation,' introduced by Robert Pagès in the late 1940s to capture the intuition that *to be is to be documentable*, based on an entity's sheer occupation of a region of spacetime, something that happens spontaneously without the need for someone else to document the phenomenon. But of course, multispecies information research takes off once human researchers are involved in documenting the activities of one or more auto-documenting nonhumans. Fuller and Lees generally agree that there is a measure of opacity between humans and nonhumans

DOI: 10.4324/9781003583424-7

in the documentation process. However, they disagree over the ethical relevance of this fact. Whereas Lees believes that animal subjects in a research setting are in principle at greater risk of exploitation than human subjects, Fuller sees the risk of exploitation as comparable if not quite equivalent.

Steve Fuller

A good way to enter the ethics of multispecies information is by considering Suzanne Briet's famous claim that an antelope becomes a document once it is placed in some classification scheme, which she understood quite broadly to include not only zoological taxonomies and nature films but also natural history museums and zoos. She seemed to suggest that these alternative forms of documentation provide different ways of accessing the animal. If we interpret this sense of 'access' in ontological terms, then the 'reality' of the antelope increases as it is subject to more documentation, especially in different media. There are a couple of ways of thinking about this prospect: One is in the spirit of 'independent corroboration,' which has been classically proposed as a criterion for establishing the existence of a novel object in science, and the other, more contemporary way is in terms of a hologram that emerges from the simultaneous projection of multiple perspectives.

This interpretation of Briet's documentary approach to the antelope fits her overall approach to librarianship, which expanded on Paul Otlet's interest in cross-classifying documents, which he regarded not as relics and residues of life but as objects possessing their own generative capacity (Fuller, 2024). But as Niloofar Solhjoo (2025) has observed, this general project—as well as the subsequent orientation of library and information science—has been explicitly 'anthropocentric,' in the strict sense of being designed to service human wants and needs. Indeed, this emphasis was arguably most pronounced in the seminal work of Otlet and Briet, both of whom saw 'documentation' as a key vehicle for making the heritage of humanity available for future *human* generations, especially in light of the 20th century's two world wars, which came close to annihilating any sense of a common humanity.

To be sure, to say that information is anthropocentric is not necessarily to disadvantage animals as co-producers of information, but it is also not to recognise them as such. This is not a trivial point, since animals have historically had a non-neutral relationship with humans. Some animals have been respected for their own qualities as well as domesticated and exploited for human purposes, whereas others have been feared, loathed and subject to systematic elimination—again relative to human purposes. So, taking seriously animals as information co-producers would require a substantial ethical reorientation. Interestingly, one of Briet's PhD students, Robert Pagès, provided one kind of solution to this problem, namely, the 'auto-document' (Day, 2024).

The basic intuition is that the existence of each individual constitutes a finite process of self-documentation. Pagès seemed to have in mind any individual being, including Napoleon's hat, though his most evocative example of auto-documentation is the living gorilla, prior to any human framing or caging. (We might call this 'Documentation Degree Zero,' with a nod to Roland Barthes.) However, to today's ears, it recalls the famous 2017 US court case that opened the door to a macaque (named 'Naruto') claiming intellectual property rights over a selfie it took with a British wildlife photographer's camera, which then figured as Wikipedia's picture of a macaque (Fuller, 2017). (The case was litigated in the US because the photographer was suing Wikipedia for unlawful use of the image, which appeared in a book he published.) This goes beyond anything that either Briet or Pagès imagined, since the judge effectively ruled that the monkey displayed sufficient competence at self-documentation to make questions of copyright relevant. However, because animals are not recognised as holders of intellectual property rights in the US, the macaque could not be declared to be the selfie's 'owner.' All that the judge could do was to deny the human photographer exclusive ownership of the image. This case suggests that at least some individual animals enjoy a *prima facie* moral right over the disposition of the information they provide, regardless of the intentions—benign or otherwise—of those who enabled it to happen and are best placed to use it.

Several issues relating to the ethics of multispecies information arise from this situation. Let me highlight two.

1 The recognition of the macaque's potential to hold intellectual property rights was hailed by PETA ('People for the Ethical Treatment of Animals') as a major advance for animal rights. This suggests that the ethics of multispecies information, while not necessarily *anthropocentric*, might need to be *anthropomorphic*, perhaps in the style of Donaldson and Kymlicka (2011). 'Rights' is a legal category constructed with mutually recognising humans in mind. Of course, historically this has not included all humans, which in turn explains the relatively late legal visibility of 'human rights.' Moreover, notwithstanding rhetorical appeals to 'posthumanism,' the most compelling examples of multispecies information transactions incorporate a healthy dose of anthropomorphism. Solhjoo (2025) refers to research in the context of animals as 'family members.' On the human side, there are concepts such as 'adoption' and 'extended family' which can make sense of the status of a dog or cat brought to live in a household. But do dogs and cats think this way? Perhaps not. More to the point, does a proper ethics of multispecies information require some sort of 'mutual recognition' between the humans and the animals about their common context—and if so, what might that look like?

2 Judgements about the ethics of multispecies information will also depend on the mode of acquisition and use of the information. In one sense, this might not seem so significant because both humans and animals acquire and use information about each other for their own purposes. The only relevant difference then may be that humans tend to proceed to Briet's more formal sense of documentation, whereas animals remain at the level of Pagès' auto-documentation. Moreover, animals trained by humans to respond in certain ways to certain environments may be seen as little different from humans educated by humans. In both cases, after this period of formal instruction, each engages freely with the other, albeit perhaps somewhat more predictably than before. The ethics of such transactions would seem to turn on whether and how 'informed consent' might be secured, especially given the background power asymmetries between humans and animals. Or could one simply concede that 'multispecies information' is ultimately an anthropocentric project whose ethical requirements go no further than showing that the animals were not harmed in the research process?

Finn Lees

The two questions you have raised—whether the ethics of multispecies information must be anthropomorphic and which modes of information acquisition might be ethically permissible—are both compelling, but before engaging with them directly, I would like to say a little bit more about Pagès' understanding of auto-documentation and its utility for understanding the case of the macaque selfie.

If a tree falls in the forest and no one hears it, can the sound of it falling be considered a *document*? I think Pagès would quite sensibly say no. Although he claimed that there exist documents 'which do not result from an imitation of an object but which are directly the object[s] responsible for providing information about themselves,' he further noted that '[w]e know, however, that such a document is not constituted as such without the addition of an auxiliary document… that integrates it into a documentary ensemble' (Pagès, 2021). His point was less to fundamentally alter our common-sense understanding of documents than to alter how we understand the process of documentation as relating to objects in the world. While traditional approaches to documentation might understand documents as existing intertextually and thus acquiring meaning only by virtue of their place within a particular documentary ensemble, Pagès pushes back against this to suggest that while a document must be integrated into a documentary ensemble, it possesses meaning before it is integrated and is therefore able to shape the ways in which its own documentation takes place. This approach certainly makes sense; as noted by Day, 'If I treat this animal as a dog, and it is a cat, then I'll know soon enough,' and so attempts to document a cat will be shaped by the information provided by the cat itself (Day, 2024: 815).

What might this tell us about the macaque? A macaque is an auto-document in the sense that it provides information about itself by looking, sounding, smelling, behaving, etc., like a macaque. Placing a macaque in an exhibition about big cats would not turn the macaque into a big cat, because the macaque itself could still exist as a document able to provide information to contradict that provided by misguided curators. The macaque that took a selfie existed as an auto-document in the sense that it provided the information that macaques are capable of taking selfies, but to be constituted as a document, the selfie still had to be situated within an ensemble of other documents. In this sense, the macaque's status as an auto-document no more implies authorship than the inherent informativeness of a tree confirms the tree as the author of works which concern it.

That copyright exists to reward authorship is only defensible—I think—on the grounds that authorship is socially useful. By allowing authors to exercise monopolies over the reproduction of their own works, they are financially incentivised both to create original works and to publish those works (Karjiker, 2014). Granting intellectual property rights to animals would not obviously incentivise the publication of animal-authored works, because animals cannot use money, and because the actual work of making such creations available for publication could not conceivably be undertaken by an animal. In part this is simply because animals are unlikely to understand what publication means, or that other people might want to see what they have created, but we might also want to agree with Marx (1974: 68–69) that animals generally produce solely in order to meet immediate needs, and that therefore while a human may edit and upload a photograph with a view to long-term gain, a macaque is only ever going to take a photograph if they feel like doing so in that particular instant.

The inappropriateness of extending property rights to animals should caution us against radical forms of anthropomorphism, which involve straightforwardly granting human rights to animals without regard for the differential capacities of different species. However, rather than insisting we reject anthropomorphism in its entirety, I would suggest that we always attempt to treat the animal as an auto-document. Whereas some animal scholars wish to adopt a postmodernist framework that sees all humans and animals as merely social constructions (Nadasdy, 2016: 9–10), Pagès' work suggests the alternative approach of seeing animals as existing outside of the socially constructed identities we assign to them. It might therefore be appropriate to treat an animal as a family member, provided that it doesn't settle the question of who that animal is. By allowing the animal to continually inform us about the ways in which their actions converge with or diverge from our normal sense of a 'family member,' this—and other categories—can be adapted and gradually brought into line with the reality of the situation.

Unfortunately, I think it is unlikely that this process would ever lead to true mutual recognition between humans and animals. Even if pets and pet owners, for example, might come to a common understanding of the nature of their familial bond, it seems unlikely that animals will ever come to understand the ways that humans gather information about them. While Hirskyj-Douglas et al. have theorised the possibility of free participation by dogs in research, whereby 'a dog would understand its choices have an impact upon the design [of research] and would therefore understand the outcomes of its choices' (Hirskyj-Douglas et al., 2015), this would only seem to work within specific ritualised encounters. Could a dog—or any other animal—ever come to understand the impact of its behaviour on human meta-discourse regarding the social status of animals? Equally, could an animal ever come to understand the ways in which its subtle bodily cues would inform a human that it needs medical treatment, let alone come to understand the specificities and outcomes of the medical treatment itself? Answering no to the latter question, Donaldson and Kymlicka suggest that 'animals are not in a position to give informed consent to treatment, and so humans must make decisions on behalf of animals, much as parents do for children' (Donaldson & Kymlicka, 2011: 143).

What I think matters then is not whether animals give their information freely, but how that information is received by humans. Do the humans in question hold rigidly to a particular conception of animality, or are they open to the informativeness of the animal as an auto-document? If the former, does their conceptual framework pose the risk of merely a factual misunderstanding, or does it lead them to actively inflict suffering on animals? In any case, might their understandings of animality be open to contestation, either from animals themselves or from other humans with different outlooks?

Steve Fuller

You have raised several issues relating to the ethical viability of multispecies information that hinge on a certain understanding of auto-documentation. I will contest this understanding, but hopefully in a way that opens discussion towards the issues facing researchers who are seeking a normatively acceptable way to characterise the multispecies information environment.

Pagès' auto-documentation presumes that the self-documenting being is what metaphysicians sometimes call a 'bare particular': Something prior to any classification (Bergmann, 1954). Its capacity for self-documentation is based simply on its spatio-temporal specificity. One can tell that it exists separately from other things in the environment—and say how that fact is registered to oneself, as another spatio-temporally positioned being. You're right that auto-documentation is observer-dependent in the sense that a document (however produced) needs to be available for others. But more importantly, Pagès is equally clear that an auto-document is not a specimen. That would already bring us to Briet's classified sense documentation. In this respect, your quote from Day about treating a dog as if it were a cat moves too fast. Sure, such things happen, but all they show is that the individual resists the 'cat' classification scheme. The logic here is not bivalent, 'either cat or dog.' The individual confirms the 'dog' scheme only if it responds in compliance with it. My more general point is that auto-documentation might be leveraged to deconstruct the default categories of documentation, which in turn might make it easier to promote a multispecies information agenda, since it would not start by presuming clear categorical distinctions between 'human,' 'animal,' and so forth.

In the case of the macaque selfie, the judge's ruling disqualified Naruto from selfie ownership simply by virtue of its nonhuman standing in the law, despite displaying a level of competence that might otherwise qualify it for ownership. If intellectual property rights were to be extended beyond humans, there would be an open question as to whether the criteria would be based on species identity or individual competence. The former would be more Briet, the latter more Pagès. If the law were to go down the auto-documentation route, it might lead to a removal of the category 'human' from being a sufficient—not only a necessary—condition for ownership. The law permits this possibility because 'rights' exist only as part of a system that includes the formal recognition of persons and their powers, liabilities, reciprocal obligations, and so forth. To be sure, John Locke is responsible for our intuition that individual humans are 'natural' persons and that everything else is either a non-person or an 'artificial' person (e.g., churches, universities, states). But prior to his intervention, legal personhood had been granted to a heterogeneous mix of beings. My point here is not to defend Naruto's right to intellectual property, but rather to highlight that the law works in such a way that, as new entities are brought into the system, the entities already in the system are expected to adjust to enable the mutual recognition of all the qualified persons. I believe that this has implications for the conduct of multispecies information research.

The most fundamental issue it raises is whether information researchers should presume the conventional species characterisation of the members of the multispecies ensemble they study. We can think about this issue in the spirit of 'grounded theory' in social research, which lets the terms of classification emerge from the research site. Of course, researchers entering the site of a zoo, a laboratory, or a household will immediately be able to draw some conventional distinctions between the human and non-human beings in their midst. However, the question for the multispecies information researcher is the extent to which such a default classification illuminates the situation. In that sense, they need to bracket Briet and return to Pagès' 'Documentation Degree Zero.'

An important reason why multispecies information researchers don't seem to trouble over what it would mean for humans and animals to arrive at a common understanding of their transactions is that information is constituted in the transaction, not in the understanding. In that sense, I share your scepticism about Hirskyj-Douglas et al.'s (2015) claim that one day dogs might come to understand the contributions they make to the research on them. However, I don't think they mean 'understand' in a deep *Verstehen*-style sense, whereby dogs come to appreciate the significance that humans attach to dog-based research, let alone how that research might benefit them as dogs. Rather, they mean—as you suggest with the word 'ritualised'—that dogs in a research setting recognise a coordination problem with humans, the regularised solution to which allows for mutual (but not the same) benefit. This is just another way of talking about how a transaction works, whether it is about setting a price for goods or establishing something as a document.

No doubt, it is ethically relevant to know how various background factors—a behaviourist might call them 'prior transactions'—have resulted in the humans and the dogs coordinating their activities in the current setting. Indeed, Robert Nozick (1974), inspired by John Locke, based his entire account of justice on this premise. Thus, if one party is being made an offer they cannot refuse, then the transaction is not free and hence unjust. Perhaps you would say that the history of prior transactions between humans and dogs since the start of domestication renders any transactions in the research environment unfree and hence unjust? Clearly, Hirskyj-Douglas et al. and other multispecies information researchers disagree, perhaps because they believe that after having been taken captive for domestication, dogs (and other similarly positioned animals) can be trained to reacquire a sense of freedom that is appropriate to the domesticated environment. In humans, we would simply call that process 'education.'

In any case, there remains a normatively interesting question about how best to characterise both the information produced in the research setting *and* the parties who produced it. For example, in the zoo situation, it might seem obvious that humans control the contexts in which information is transacted with animals (e.g., feeding, watering and more generally ensuring their security), as well as fix the identities of the animals as members of various species. However, in the spirit of auto-documentation, the multispecies information researcher might wish to focus on the nonhuman resistances to these presumptive characterisations, without automatically shifting to a narrative of human species domination, whereby class warfare is reinscribed as species warfare. A better starting point might be Gregory Bateson's (1979) definition of information as 'the difference that makes a difference,' which is rooted in William James' pragmatist theory of meaning and Shannon and Weaver's mathematical communication theory. In short, the identities of both sides of the transaction emerge in the course of observing how the resistance is played out.

Part of the craft of research in this setting involves arriving at an appropriate way to name the parties on both sides of the resistance. For example, Wolfgang Köhler's (1925) classic work on ape intelligence is bookended by general remarks about the psychology of chimpanzees, but the bulk of the text is devoted to following the problem-solving styles and strategies of several chimpanzees who are given human names and personalities. Much to the consternation of behaviourists, not only was Köhler deliberate in his anthropomorphism, but he also justified it in terms of the common evolutionary heritage of humans and apes. He regarded his chimpanzee subjects as humans with a somewhat greater communication barrier. His term of choice was 'anthropoid.' (One might think of Jane Goodall as an extender of this approach.) And to a large extent, Köhler succeeded, insofar as his work was cited by researchers in *both* human and animal intelligence for much of the 20th century, even though his book is really just about apes. I believe that he managed this feat by focusing on the individuality of his subjects to such an extent that he effectively blurred the boundary between human and ape at the narrative level.

Here, it is worth recalling that the post-Darwin drive to reduce the difference between humans and animals came from two directions, one towards less and the other towards more anthropomorphism. Behaviourism, the standard-bearer for the first approach, favoured mass (statistical) studies of organisms in controlled settings with limited protocols for engagement in information production. Behaviourists drew a sharp distinction between studies in the laboratory and in the wild, typically using the lab to validate observations in the field—not the other way round. Köhler exemplified the second, more 'ethological' approach, which operates by 'field experiments,' somewhat in the spirit of ethnomethodologists in sociology, effectively blurring the distinction between the lab and the wild, resulting in a blended, more qualitative account of multispecies information. Indeed, Köhler did not hide that he entered his research already thinking that he and the apes share a common intelligence but express it differently. And here I would add that to presume a common intelligence is not to presume a common understanding. (In Saussure's terms, it's the *langue/parole* distinction; in Chomsky's, it's competence/performance.)

Now, I'm not arguing that multispecies information researchers need to follow Köhler's example to the letter. However, the spirit of his enterprise involved bracketing presumed species differences as determining the character of human–ape transactions. Thus, he was sensitive to the extent to which humans treated the apes as they would other humans, and similarly, apes dealt with humans as they would other apes. The methodological point is especially salient in the context of humans and animals in a common household, where it is common for people to refer to animals as 'family members,' often treating the information transacted with them at least as seriously as they would with human household members. In this respect, the multispecies information researcher is well placed to challenge stereotypes about species differences that arise from a specimen-based approach to animal behaviour studies, in the spirit of Pagès' concept of 'auto-documentation.'

Finn Lees

Our disagreements regarding how to interpret Pagès are, I think, minor. Sure, an auto-document is not a specimen because it is not 'a token of a type' (Day, 2024: 814), and as such, an individual auto-document does not necessarily demand to be treated as a representative of a particular category. However, it does not follow that the quote about treating a cat as a dog moves too fast. Most people who are aware of dogs are also aware of cats, and so if a cat resists the classificatory scheme of 'dog,' then it will not be long until its classifier moves on to propose 'cat' as an alternative classification. Of course, as auto-documents are particulars, not specimens, any individual cat will, to a greater or lesser extent, provide some resistance even to its classification as a cat, and anyone wishing to cultivate a close relationship with that cat will have to get to know it as an individual rather than merely as the feline archetype. In other words, they will have to be open to the informativeness of the animal as an auto-document.

On an individual level, that is generally how things play out. First, we learn in general terms how to interact with cats, dogs, and other species, and then we learn the specific personality quirks of those individuals we see on a regular basis. However, to treat a dog as a dog—or to treat a particular dog in a manner befitting them as an individual—would mean one thing in Yorkshire and another in Yulin. The ethical treatment of animals is not only about learning how to behave towards them, but also the moral categories through which they are understood. Here, you would be right to point out that logic is not bivalent. If animals resist classification as automata, does that make them people? Not only are there a nearly infinite number of ethical categories through which animals might be understood, but it is also not always clear what it would mean for an animal to resist a particular category.

You have suggested that the benefit of the concept of auto-documentation is that it might allow us to avoid the categories of 'human' and 'animal' and thus regress to the pre-modern status quo of granting legal personhood to a range of different types of entity. While I am not against extending the moral category of personhood to animals, doing this would not be to embrace auto-documentation in the sense that you have argued, as personhood would still be a classification imposed on animals from without. There is perhaps a sense in which legal and political assumptions of human personhood form the basis for *humans* to evade classification, but this cannot be disentangled from the social processes which have rendered pre-modern notions of personhood untenable.

Mediaeval law was able to accommodate the existence of nonhuman persons in large part because mediaeval society made no pretence of equality of persons. There was no need for a society to deny personhood to pigs based on their ontological difference from humans when that same society understood different humans (men and women; serfs and kings) to be just as fundamentally unequal. Various Enlightenment thinkers wrote theoretical denunciations of that social order, but ultimately it was overthrown because the broad masses of the people were able to demand equal treatment. We don't just owe the intuition that individual humans are natural persons to John Locke—we also owe it to the Jacobins, the Haitian slaves, and the Suffragettes. Society now is—at least in theory—founded upon the equality of people, which is made possible by human participation in collective social life.

In such a society of equals, it might be possible to 'deconstruct default categories,' as you put it, both as these categories relate to documentation and to other aspects of interpersonal conduct. John Stuart Mill, for example, argued against the separate legal treatment of women on the basis that 'nothing final can be known' about gender differences prior to the free development of women as independent subjects (Mill, 1878: 44). In a similar spirit, many later feminists have argued for the deconstruction of gender altogether as a social and biological system of classification, an approach which might require us to treat fellow humans as nothing more than bare particulars, outside of any imposed system of classification (Dworkin, 1974: 174–193; Rothblatt, 1995). Such approaches to gender, however, are only possible because of the capacity of humans for self-creation. Humans are not only beings for whom *being* is an issue (Heidegger, 1962: 32), but they can also understand the identities assigned to them by others within society—and conform to or rebel against them, as the case may be (Butler, 1990; Goffman, 1956).

I think it is fair to say that in almost all instances, animals lack a comparable capacity for self-creation. This would seem to be a corollary of your suggestion that animals might recognise coordination problems with humans without understanding what interspecies interactions mean to humans. Under these circumstances, an animal would be able to intentionally behave in a manner likely to elicit a particular reaction, but it would have no means to knowingly contest the identity imputed to it. As such, whether or not humans have complete control over the contexts in which they interact with animals, they will have complete control over how they categorise the animals with which they interact. According to my reading of Pagès, this would be compatible with auto-documentation, which I think requires only that classifications be responsive to disconfirming evidence, but according to your reading, I think it would not, as the slightly stronger definition that you adhere to requires that classification not be attempted in the first place. While it is theoretically possible that a human could refuse to categorise an animal according to any system of classification, this would undermine the possibility of human action with respect to that animal. In knowing how to interact with a dog, for example, it might be necessary to have some basic idea of what dogs are and how they differ from other species, but it would also be necessary to decide whether the dog is a person, an automaton, a source of meat, etc. Without such a decision, it would be doubtful that an interspecies interaction could be orientated towards a particular goal or that it could be constrained by any ethical concerns.

A few weeks ago, I attended a protest against the use of beagles in experiments. We drove to various businesses that work with animal testing labs and set up outside with placards and loudspeakers to announce their complicity in puppy murder. Whenever employees of the companies made themselves visible, several of my companions took it upon themselves to engage them directly, repeatedly shouting the questions, 'Do you have pet dogs?,' 'Do your colleagues have pet dogs?,' and 'Will you go home to your dogs this evening?' These were not calls to deconstruct the categories through which dogs are understood but rather an insistence that the category of *pet*, which perhaps also implies personhood and family membership, provides a framework for understanding dogs in general. Crucially, this demand was made by protestors who possessed no direct knowledge of the dogs on whose behalf they were advocating, which meant that they could not be plausibly seen as being responsive to the dogs as particular entities.

Where I hope we might agree, however, is that there is something profoundly progressive about the treatment of animals as potential members of a family. You have written that when animals are treated as family members, the information they provide can be taken as seriously as that provided by humans, and that this can be a basis for undermining species stereotypes. I would add that the human–pet relationship is motivated principally by love, and that an openness to the animal as an auto-document, in the sense that I understand, is therefore encouraged. To paraphrase Donna Haraway, the ideal pet owner '[knows] more at the end of the day than they did in the morning, because they owe it to their beloved' (Haraway, 2008: 107). Here, pet ownership can be seen to stand in contrast to the highly instrumental ways of relating to animals prevalent in animal testing and animal agriculture, which are interested in the animal as a biological model and a producer of commodities and so develop technologies for standardising animal life (Blanchette, 2020). Those who truly love animals cannot avoid categorising them, but they cannot impose standards on them either. Rather, they must be receptive to the information provided by the animal regarding its individual particularities for the purpose of learning to '[allow] the other to be' (Solhjoo, 2024: 12).

Conclusion

Lees' final response highlights the ethically relevant tension running through this dialogue. While he embraces the idea of animals as 'family members,' he also thinks that humans need to allow them 'to be.' In Fuller's terms, this looks like a strong endorsement of anthropomorphism, including provision of space for the animal to develop its own personality in the human-based environment. It comes close to how humans normally regard their children in the modern family setting—except that Lees focuses on the learning experience of the pet owners *qua* surrogate parents, instead of the animals *qua* surrogate children, to whose minds he does not pretend to have access. Perhaps this helps to explain his concluding appeal to *love* as the ethical stance towards animals in information research, especially if we think of love as 'unconditional,' that is, without requiring a proportionate—if any—response from the loved one. In that sense, Lees may be suggesting that regardless of why we might wish to acquire information from animals, the effect of the information transactions should leave the animals in roughly the same or better state to remain the source of our love.

There are two ways to think about this matter, one practical and the other theoretical.

On the one hand, the multispecies information researcher may do whatever it takes to keep the relationship with the animal on good terms, which includes suspending any attempt to understand how the animal processes the information transaction. It may even mean that the researcher and the animal treat the means-ends character of their transaction in reverse terms, whereby, say, the human provides food to the animal in return for information, while the animal provides information in return for food. If we presume the level of interspecies cognitive opacity that Lees does, then no 'deception' is involved, since neither party knows what the other needs to know and hence cannot be accused of failing to disclose it. It is also compatible with Lees' love-based standpoint, which does not require a specific response from the animal but only the researcher's understanding that the animal's response is consistent with the love that the researcher is trying to show towards the animal.

On the other hand, and closer to Fuller's view, the multispecies information researcher may reason that their own and the animal's capacity for interspecies understanding overlaps sufficiently that a more robust epistemic and ethical incorporation of the animal into the human lifeworld is possible, in the manner of Köhler (1925) or even Donaldson and Kymlicka (2011). Whereas in the early 19th century, Friedrich Schleiermacher famously proposed 'empathy' as the feeling that enables humans to reach out beyond themselves to recognise each other's shared divine lineage, by the end of that century, Wilhelm Dilthey had replaced this theological justification with a biological one based on Darwin's views on the evolutionary unity of all life from a common primitive ancestor, though Dilthey continued to maintain that humans differed from animals by virtue of the degree of control that they can exert over themselves and their environment (Bulhof, 1980: Chapters 6–7). More recent and in-depth studies of animal minds across many species have reduced the interspecies gap still further, pointing to a time when anthropomorphism might eventually be placed on a sound scientific footing (Griffin, 2001). In any case, this trajectory suggests that the ethical horizons of multispecies information research have only begun to be explored.

References

Bateson, G. (1979) *Mind and Nature: A Necessary Unity*. New York: E.P. Dutton.

Bergmann, G. (1954) *The Metaphysics of Logical Positivism*. New York: Longmans.

Blanchette, A. (2020) *Porkopolis: American Animality, Standardised Life, & the Factory Farm*. Durham, NC: Duke University Press.

Bulhof, I. (1980) *Wilhelm Dilthey: A Hermeneutic Approach to the Study of History of Culture*. The Hague, NL: Martinus-Nijhoff.

Butler, J. (1990) *Gender Trouble: Feminism and the Subversion of Identity*. New York: Routledge.

Day, R. (2024) Robert Pagès' concept of the "auto-document" as a forerunner to neo-documentation's philosophy of documentality. *Journal of Documentation* 80: 813–823.

Donaldson, S. and Kymlicka, W. (2011) *Zoopolis: A Political Theory of Animal Rights*. Oxford: Oxford University Press.

Dworkin, A. (1974) *Woman Hating*. New York: E.P. Dutton.

Fuller, S. (2017) Monkey selfie case settled – but there are many similar animal rights battles to come. *The Conversation*, 18 September. https://theconversation.com/monkey-selfie-case-finally-settled-but-there-are-many-similar-animal-rights-battles-to-come-84000

Fuller, S. (2024) The fate of the document in library and information science. *Library Trends* 73: 64–79.

Goffman, E. (1956) *The Presentation of Self in Everyday Life*. Edinburgh: University of Edinburgh Social Sciences Research Centre.

Griffin, D. (2001) *Animal Minds: Beyond Cognition to Consciousness*. 2nd ed. (Orig. 1992). Chicago, IL: University of Chicago Press.

Haraway, D. (2008) *When Species Meet*. Minneapolis: University of Minnesota Press.

Heidegger, M. (1962) *Being and Time* (trans. J. Macquarrie & E. Robinson). Oxford: Blackwell.

Hirskyj-Douglas, I., Read, J. and Cassidy, B. (2015) The doggy ladder of participation. In *Proceedings of British Human-Computer Interaction Conference*. Lincoln (July). https://www.researchgate.net/publication/280131453_Doggy_Ladder_of_Participation_DLOP

Karjiker, S. (2014) Justifications for copyright: The economic justification. *South African Intellectual Property Law Journal* 2: 13–41.

Köhler, W. (1925) *The Mentality of Apes*. (Orig. 1917). New York: Harcourt Brace and World.

Marx, K. (1974) *Economic and Philosophic Manuscripts of 1844* (trans. Martin Milligan). Moscow: Progress Publishers.

Mill, J. S. (1878) *On the Subjection of Women*. London: Longmans, Green, Reader, & Dyer.

Nadasdy, P. (2016) First nations, citizenship and animals, or why northern indigenous people might not want to live in Zoopolis. *Canadian Journal of Political Science* 49(1): 1–20.

Nozick, R. (1974) *Anarchy, State and Utopia*. New York: Basic Books.

Pagès, R. (2021) Transformations documentaires et milieu culturel. (Orig. 1948). *Proceedings from the Document Academy* 8(1): 3.

Rothblatt, M. (1995) *The Apartheid of Sex: A Manifesto on the Freedom of Gender*. New York: Crown Publishers.

Solhjoo, N. (2024) The red thread of information across more-than-human love. *Journal of Librarianship and Information Science* 1–16. https://doi.org/10.1177/09610006241255418

Solhjoo, N. (2025) The multispecies perspective in library and information science. *Information Research* 30: 323–330.

7

RESEARCH PATHWAYS FOR MULTISPECIES INFORMATION SCIENCE

Niloofar Solhjoo

Doing Multispecies Research in Information Science

Multispecies inquiry is an interdisciplinary research approach grounded in anthropology, science and technology studies (STS), and critical animal studies. It aims to deepen ethical, political, and epistemological understanding of the diverse forms of life entangled in relationships of knowing and living together.

This final chapter in Section 1: *Fundamentals*, before moving into Section 2: *Studies*, explores how multispecies inquiry can inform Library and Information Science (LIS) research and practice. It offers epistemological and methodological approaches while also acknowledging the limitations we face as humans when attempting to engage with other species. I draw on ethical and practical insights from my experiences researching with animals in LIS, as well as from multispecies scholarship beyond the discipline, which is developing across the humanities and social sciences. Encouraging deeper connections with animals, you are invited to walk the edges of human-centred research methods in LIS.

Being Human

Multispecies research in LIS starts with a foundational question: What does it mean to be human? The framework of humanism has long emphasised human superiority. The problem is not humanity itself but the human-centred perspective that shapes our interpretations of the world (Haraway, 2003). This has reinforced anthropocentric assumptions about what counts as information, documentation, or knowledge. It has also shaped who qualifies as an information user, information provider, or knowledge creator—and who is entitled to participate in decision-making.

DOI: 10.4324/9781003583424-8

All research and practice in LIS is (and will always be) conducted from a human position. Emphasising the relational networks among all beings, the multispecies turn does not seek to erase the human but to decentre it (van Dooren et al., 2016). We therefore enter the multispecies context with an inherent constraint: The limitations of humanism itself.

This turn acknowledges the limitations of human understanding and control. It resonates with Wolfe's (2010) notion of shared finitude: Humans and nonhumans alike are vulnerable, embodied, and mortal. Both are shaped by evolutionary processes and bound by interwoven destinies (Dunn, 2021). Multispecies research is attuned to assemblages and relationality across species (including humans). Multispecies research attends to assemblages and relationality across species, including humans. Several chapters in this volume illustrate these entanglements—for example, in information pedagogy (Solhjoo & Hartel, 2026, Chapter 5) or through the concept of *distributed agency* in information practice (Ekström & Pilerot, 2026, Chapter 12).

Multispecies research also recognises the autonomous agency of entities (e.g., geological processes) that exist beyond human relationality (Bonelli & Walford, 2021). These presences act independently of human perception or control, a perspective echoed in Chapters 4 and 14 of this volume. Ronald Day highlights the *powerful particulars* embedded in documentation, while Erik Radio investigates the vast, often invisible presences that function as *massive unseen companions* within document studies.

These strands of thought underline a core challenge: We cannot fully access the subjective experience of another being. This limitation is not only philosophical but also epistemological, influencing how we study other species and understand their lives (and embedded information).

Animal Inclusion

Even though we cannot fully escape our human bodies and perceptions, we must not overlook the connections that link us with other species (Bonelli & Walford, 2021; Wolfe, 2010). Haraway (2016) suggests that instead of trying to solve or escape our limitations, we should view research as an ethical task of *staying with the trouble*. The word 'trouble' originally meant to stir or disturb something, and Haraway uses it to describe the need to remain fully present with the complex and messy realities of life today—such as staying with the animals in LIS. This means making our research and practices open to living and thinking with others as *companion species* (Haraway, 2003). It calls us to live as mortal beings entangled in unfinished and evolving shared lives, worlds, and meanings.

In my own doctoral research exploring information experience within multispecies families (Solhjoo, 2024a), I employed an approach that conceptualises the human–animal couple in their natural home environment. The objective was to understand the relational 'red thread' of information (Bates, 1999) that shapes daily life with companion cats and dogs. The research moves beyond the classic model in which a human gains information either from the environment (consciousness) or from another human (communication). Instead, it proposes a new model in which both humans and animals are active agents, each perceiving their environment and engaging with one another, collectively shaping shared information experiences.

A major discussion point was the methodological challenge of studying a cat's or dog's subjective experience without a shared language and how to conduct research with them in a participatory form. I adopted a phenomenological approach, using ethnographic methods including observation, photovoice, and sensory walks (Solhjoo et al., 2022). I also reflect on other methodological challenges of focusing on animals in research conducted in natural environments. These include ensuring voluntary participation and consent from animals, the validity of data gathered from animals, and how findings are ethically represented (Solhjoo, 2024a).

To address these challenges, I drew from the field of Animal–Computer Interaction (ACI), which has developed methods for designing with and for animals. ACI offers ongoing discussions about prioritising animal interests over human ones. For example, the 'Doggy Ladder of Participation' (Hirskyj-Douglas et al., 2015) is a model that outlines different levels of a dog's involvement in research. This framework helped me embed principles such as freedom, informed choice, and empowerment for cats and dogs into my information experience research (see also Chapter 3; Gorichanaz (2026) builds on ACI to expand multispecies information experience).

In my study, informed consent for animal participants was interpreted through continuous observation of body language. The animals retained full autonomy and could withdraw from the research at any time. They were free to act naturally in their homes, choosing either to remain under camera observation or to leave the space altogether. I also committed to representing the animals' perspectives in the analysis and research outcomes by focusing on their presence, interactions with the surroundings, and tuning into animals' emotional states. Their perspectives highlight non-verbal yet novel forms of information in everyday life information behaviour studies. The ethical discussion surrounding multispecies information science is further explored in Chapter 6 by Fuller and Lees (2026).

Throughout a multispecies information science project, your goal is to create ethically sound research and practice while acknowledging your significant limitations. This requires what Wolfe (2010) calls *reflexive awareness*: Constantly questioning how your own assumptions and frameworks shape both your understanding of the animals and your relationship with them. Accepting inherent uncertainties is essential for conducting responsible and meaningful research that respects the animals involved.

Ways of Knowing

To bring nonhuman animals into Information Science research, I outline two major epistemological orientations that form the basis of many methodological approaches in multispecies research. I refer to them as the *between-beings* and *within-the-being* lenses.

- **The between-beings (outer lens)**: This lens situates all species, including humans, within broader ecological, cultural, social, or technological systems. It emphasises relationality and entanglements, examining how animals, plants, and landscapes, and even non-living agents co-constitute information, knowledge, and understanding. A critical concept here is Donna Haraway's (2003) notion of *companion species*, which blurs the ontological boundaries between human and nonhuman. This relationship is reciprocal; both humans and others are mutually shaped through their interactions. True understanding of another species and their universe of information requires immersion into their world through mutual relations and shared representations (Bonelli & Walford, 2021; Smith & Mitchell, 2012). Methodologically, researchers combine evolutionary biology, ethology, and ethnographic and phenomenological perspectives to study human–nonhuman couples and entanglements, using concepts like assemblage and intra-action.
- **The within-the-being (inner lens)**: It focuses on the innate, biological, or phenomenological dimensions of life. It prioritises understanding the embodied mind of a species on its own terms, seeking to answer: How does this being perceive, process, and generate information within its unique umwelt (subjective perceptual world) (Uexküll, 2010)? Methodologically, there are objective, data-driven methods like ethology and more interpretivist, social approaches like ethnography, which narrates the life and experiences of an individual creature. While this inner lens is essential, it also underscores a major challenge: The difficulty of accessing the subjective experience of another species. It is like trying to explain sight to someone blind since birth. Such efforts are inherently limited, as they sometimes rely on imaginative projection (Nagel, 1974).

You can find an example of these complementary orientations in multispecies information research in Chapter 2, where Bates (2026) elaborates on the study of multispecies information behaviour through the lens of ecological psychology. She suggests that, in conducting such research, one should describe the characteristics of the umwelts of the species involved (an *inner lens* that immerses the researcher in the species' ways of being and doing), as well as characterise the physical ecology and its affordances for those species (an *outer lens* that focuses on the environmental and relational context).

Another example comes from the phenomenological structure of documental becoming outlined by Gorichanaz and Latham (2016). They argue that a document emerges through the dynamic relationship between a person and an object, as meaning is created through transactions (of different forms of internal and external information) between person and object. This perspective allows for imagining other types of documents, such as human–animal, animal–object, and animal–animal. Their documentality operates at all levels, from the smallest parts of a document to the broader systems in which it is embedded. Importantly, this relationship is both fractal and recursive: The document system provides the external context that gives meaning to each part, while the cumulative experience of the parts shapes the meaning of the whole. This creates a holistic framework in which the inside (parts) and the outside (systems) are mutually constitutive; each shapes the other.

The crucial point in designing multispecies information research is that neither lens alone is sufficient. The inner lens offers an embodied, biological, and empathic entry point, while the outer lens maps the shared ground where attunement (tuning into animals' worlds) and co-becoming occur.

In the following sections, I discuss both interpretive and scientific approaches that can support the within-the-being and between-beings lenses in multispecies information research. The way forward is to recognise the limitations of each methodology while creatively integrating its respective strengths.

Animal Phenomenology

Phenomenology offers philosophically grounded methods for accessing the lived experience (Käufer & Chemero, 2015). It is essential for studying how beings encounter and are shaped by information (Gorichanaz, 2020). It shifts the focus from a third-person perspective to a first-person perspective in information studies and enables researchers to explore how information is lived and experienced from within a (human or nonhuman) person's consciousness (Gorichanaz, 2018).

While traditionally focused on human experiences, phenomenology can extend to exploring the lifeworlds of nonhuman animals (Lestel et al., 2014; Lewis & Owen, 2020; San Martín & Peñaranda, 2001). Animals are conscious, intentional beings capable of interpreting and responding to their environments. Animal phenomenology asks the fundamental question: What is it like to be a specific animal?

The goal of using phenomenology in multispecies research is not to gain direct access to animal experience (which is ultimately impossible) but to help form a relational partnership with other species (i.e., to create a human–animal couple) that fosters shared understanding. Shores (2017) provides a useful category of strategies for this purpose, which are briefly explained below:

- Translational strategy using communication and empathy: This approach involves engaging in two-way communication with species capable of interacting with humans (e.g., dogs using sign language, or plants through bioelectrical signals). By observing an animal's body and behaviour, and using empathy, we try to interpret their emotions through our own physical and emotional experiences. While this strategy can be insightful, it has clear limitations. It relies on shared emotions, which may not exist across all species. Moreover, our communicative tools are often ineffective or unavailable for many species, making this strategy limited.
- Transpositional strategy from Uexküll's umwelt theory: Think of an animal's Umwelt as its personal reality bubble. This bubble contains everything it can sense and everything it can do (Uexküll, 1934). Scientists use experiments to identify which stimuli (colours, sounds, smells, etc.) an animal responds to and map the animal's Umwelt: What it senses and how it behaves. We can then try to transpose ourselves into that structured model by focusing only on the senses we share with the animal. This strategy only grants access to the perceptual side of the animal's world, not its inner experience or intentions.
- Transformational strategy from Deleuze & Guattari's becoming-animal: Becoming-animal is not about literally turning into an animal or imitating one. Instead, it is about forming a deep connection or alliance with another species by sharing its effects. It involves physically and relationally restructuring one's own body and behaviours to approximate the affective and interactive capacities of an animal (Deleuze & Guattari, 1980). To become-animal, you form a group with an animal and change your own body's relationships and behaviours to align with the actions of that animal. This may involve shifting your focus away from the visual landscape to other senses. The goal is not to perfectly copy the animal but to enter a zone of proximity where you and the animal form a new, hybrid entity.

In Information Science, phenomenological methodologies are increasingly used to explore assumptions about knowledge, information, and human–information experience (Budd, 2005; Gorichanaz, 2020; Suorsa, 2024; VanScoy & Evenstad, 2015). Phenomenology offers not only methodological insights but also ontological and epistemological critiques of longstanding LIS concepts and paradigms. For instance, the work of Bruce et al. (2014) and Gorichanaz (2020) has been instrumental in introducing information experience as a phenomenological approach to LIS, with particular emphasis on *understanding* as a central epistemological concept. Understanding is viewed as a lived, interpretive, and hermeneutic process, embedded in embodied and contextually situated forms of information (Gorichanaz, 2020). Through this lens, information becomes something we engage with existentially (i.e., part of who we are and how we become in the world), thus calling for a relational view of information rather than an objectified one.

The experiential lens of information (as a strand of phenomenology in LIS) helps me understand nonhuman information experience and contributes to the development of multispecies information science. It aligns with an a posteriori orientation, attending to the experience of being informed by anything that an individual regards as informative (Yu & Liu, 2022, p. 10). It is attuned to diverse aspects of an individual's relationship with information, ranging from experiencing sensory information to cognitive–affective information, and from acquiring to interpreting it (highlighting both inner senses and perceptions as well as outer relational and environmental entanglements) (Savolainen, 2019). Finally, this approach opens the possibility of expanding the first-person experiential focus in LIS to include nonhuman agents as individuals capable of creating, seeking, using, and interpreting information (Solhjoo et al., 2022).

So, I combine information experience, as a phenomenological approach, with multispecies ethnographic methods to explore information that shapes understanding between humans and animals. It helps me move beyond an anthropocentric view and see understanding as an ongoing, environmentally distributed process shaped by the assemblages of multiple agents living in a house (Solhjoo, 2024b).

There are many other phenomenological approaches for multispecies information science. Among them are the ecological–phenomenological perspective (Dashper & Brymer, 2019) and performance phenomenology (Lewis & Owen, 2020). While these approaches have been applied within LIS, primarily in human-centred contexts, they have been used in other fields to advance multispecies relationships in their theory and practice. Researchers adopt these approaches to recognise animals as intentional, responsive, and interpretive agents, embedded in meaningful relationships with their environments.

The performative lens in phenomenology emphasises that we can approach animal experience through embodied performance, temporarily reconfiguring our own sensory and bodily apparatus to approximate how other species live. This involves not just observing animals but physically and sensorially engaging with their lifeworld. Suorsa (2015) applies this insight in LIS by using the phenomenology of play and embodiment to conceptualise the human–information relationship as an ongoing, dialogical event, rather than a one-way processing of data. Her work situates information within the context of lived, ecological, and ethical entanglements, calling for a broader sustainability ethic in LIS (Suorsa, 2024). Outside LIS, Foster (2016) uses performance phenomenology as embodied imitation and sensory reorientation to inhabit the perceptual world of animals. He combines animal physiological research with immersive fieldwork, adapting his behaviour, posture, and movement to reflect the animal's subjective world.

As a researcher using performance phenomenology, empirical data collection techniques such as audio recording, video recording, and field note-taking are essential to clearly document all actions. Reflective techniques are also important for noticing shifts in your own bodily awareness, emotions, and perceptions as you attune to other species while engaging with them through playful or imitative practices. Moreover, a performance-based approach in animal phenomenology requires more than imitation. You need to conduct background research on the species' physiology, sensory capabilities, behaviour, and ecology by consulting literature in the natural sciences (Lewis & Owen, 2020).

The ecological lens in phenomenology, drawing from ecological psychology, shifts attention to the animal–environment system rather than focusing on the inner life or isolated behaviour of a species. This approach treats all animals, including humans, as deeply embedded in evolving systems of mutual influence. Two key concepts within this framework are constraints and affordances, which are factors that limit, enable, or channel behaviour. Together, constraints and affordances allow researchers to explore how meaning and behaviour emerge within the specific ecological and bodily context of nonhuman lifeworlds (Dashper & Brymer, 2019). In LIS, there are relevant examples. A recent one is the critique of the Data–Information–Knowledge–Wisdom (DIKW) pyramid by Calzati (2025), who argues that it represents an overly linear and anthropocentric model. Instead, he proposes an ecosystemic view rooted in the concepts of affordance and constraint. You can find another example in this book where Bates (2026) discusses how to study humans and animals in common ecologies. From this perspective, information (as well as knowledge and data) is not a static entity but a relational and emergent process shaped by embodied perception and environmental interaction.

By combining experiential, performance, or ecological perspectives with animal phenomenology, a powerful framework emerges for multispecies information research. This approach moves beyond the detached observation typical of the natural sciences, enabling researchers to step into the perceptual and experiential edges of other beings' worlds. Operationalising these phenomenological strategies involves a range of methods, from scientific (e.g., experiments) to interpretive ones (e.g., ethnography), as discussed in the following section.

More-Than-Human Participatory Research

In LIS, participatory research, such as photovoice, photo diaries, and participant-generated drawings, is an established method (Given et al., 2011; Greyson, 2013; Pollak, 2017) in which research participants are not just subjects of study but active contributors or co-researchers. Participatory methods typically involve the researcher commissioning a task (e.g., creating images or making activities) as part of the research project, with the task assigned to a community of participants, with or without the collaboration of the researcher. Informants are provided with the necessary tools and instructions and are asked to participate in aspects of the subject under study within their own environment. These methods generally capture the participants' contextual perspectives and allow the researcher to metaphorically see through the participants' eyes.

More-than-human participatory research builds on these foundations but extends participation beyond human subjects to include interactions with non-human beings and materials. This perspective emphasises the co-creation of knowledge through entanglements between humans, other species, and the material environment (Bastian et al., 2017; Mancini & Lehtonen, 2018).

It does not suggest that animals or ecosystems literally communicate in human terms (Noorani & Brigstocke, 2018). In this context, participation goes beyond intentional speech or action. It includes shared temporalities, embodied practices, and affective exchanges that contribute to the shaping of shared worlds and are often facilitated by mediating artefacts (Lupton, 2019).

To put this into practice, for example, the next time you walk your dog, try letting the dog lead the way. Pay close attention to where it sniffs, pauses, or urinates—these behaviours can reveal dimensions of your daily dog walk that might otherwise go unnoticed to you. In this way, your dog becomes your co-researcher. Your dog's sensory experiences, movements, and interactions with the environment uncover hidden layers of your neighbourhood, expanding what you can feel and understand.

In my own exploration of information experience within multispecies families (Solhjoo, 2024a), I found that once you begin working with human–cat or human–dog assemblages, you naturally become more attuned to many other things and start incorporating them into your fieldwork. A cricket caught by a cat participant and offered as a gift to its human, a horse encountered by a dog participant on a beach walk, and a neighbour met by another cat participant during its wanderings all contributed to helping me achieve my research objectives.

In biodesign, an interdisciplinary practice that incorporates biological systems, organisms, or processes into design, Kim et al. (2024) employ virtual reality to simulate microbial behaviours, such as motility and inter-organism communication. The aim is to help designers develop sensitivity towards microbial life and engage with these organisms as meaningful, albeit invisible, collaborators in the design process. This reorientation towards multispecies co-design also appears in ACI, which evolves with the animal's interests and prioritises their welfare and capacity for expression in interspecies systems (van der Linden, 2022; Westerlaken & Gualeni, 2016).

More-than-human research can also enhance human–human interaction and deepen our understanding of our own species. For instance, social animals such as dogs often act as social catalysts, making conversations during fieldwork more accessible and natural (Vincent, 2019). This is one reason why species like cats and dogs have already been integrated into library services for their users (see Chapter 5 by Solhjoo and Hartel on the multispecies theme in libraries and among librarians).

Additionally, this approach disrupts conventional knowledge hierarchies by drawing on diverse ways of knowing. It embraces not only scientific approaches but also Indigenous, decolonial, and alternative epistemologies that have long recognised the agency of animals, plants, rocks, and ecological forces (Noorani & Brigstocke, 2018; Tarcan et al., 2022; Van Dooren et al., 2016). This recognition stems from the Indigenous worldview that always assumes a profound sameness, and therefore a sense of recognition, between the abilities and sensibilities of objects and those of humans (Jones & Hoskins, 2016).

In recent decades, various theoretical frameworks have emphasised socio-technical relations, with Actor-Network Theory (ANT) being one of the most influential in more-than-human participatory research (Nimmo, 2011; Westerlaken & Gualeni, 2016). Developed by Bruno Latour and colleagues, ANT provides a lens for understanding how knowledge, practices, and relationships emerge from networks of human and nonhuman actors (Latour, 2007). For instance, in LIS, ANT has been used to analyse how nonhuman actors, such as e-learning objects, shape information literacy (Schreiber, 2017). In ACI, it has been used to show how animal interactions with technology can co-create meaning (Westerlaken & Gualeni, 2016). Despite its strengths, ANT's ontology has been critiqued for failing to distinguish between sentient and non-sentient actors, which can obscure ethical responsibility (Cohen & Cohen, 2017; Dryzek, 2016). However, when combined with other multispecies approaches, ANT remains a valuable tool for tracing more-than-human practices.

Multispecies Ethnography

Ethnography is fundamentally the task of producing descriptive, engaging, and story-like accounts about a particular culture (Spradley & McCurdy, 1972). It is grounded in key principles such as naturalism, holism, contextual understanding, the use of both emic and etic perspectives, and a non-judgemental approach (Fetterman, 1998). Central to ethnography is immersive fieldwork, which typically involves a combination of methods including observation, fieldnotes, interviews, and the analysis of documents (Bernard, 2018).

Over time, ethnography has evolved to include a range of specialised subfields or ethnographic variations. These adaptations reflect shifting cultural and academic landscapes and contribute new tools, techniques, and domain-specific knowledge to the broader ethnographic tradition (Hartel et al., 2025). Among these, multispecies ethnography has gained prominence over the past decade, following the influential work of Kirksey and Helmreich (2010). It studies the relationships and interactions between human and non-human species as they shape culture, politics, and ecosystems. It rejects the idea that only humans have culture and agency and argues that nonhuman actors (animals, plants, and fungi) are active participants in shaping the world.

Multispecies ethnographers strive to create a language of togetherness, emphasising Haraway's (2003) concept of companion species, which suggests that being a subject requires relationships and that understanding emerges through living and engaging together, rather than merely observing or probing. So, the data collection and analysis in multispecies ethnography seek to engage with the assemblages of agentive beings and their interrelations in a particular context (Ogden et al., 2013).

Rooted primarily in interpretive approaches, these methods involve *listening with* nonhuman species. Ethnographers from various social science fields (e.g., anthropology, education, geography, urban design, psychology, health studies, leisure studies, and social work) have developed innovative methods to responsibly represent the perspectives and experiences of nonhuman participants in their fieldwork. A recent event that centred on these issues was the *Exploring Human–Animal and Multispecies Relations: Risk Taking in Research Methods Symposium*, held at the Centre for Arts and Wellbeing, University of Brighton, in June 2024. This symposium brought together scholars from diverse disciplines to explore creative and qualitative approaches to researching human–animal and multispecies entanglements. The event featured lectures, discussions, and artistic presentations that challenged anthropocentric methodologies and foregrounded nonhuman agency, sensory experience, and embodied encounters across varied contexts.

Multispecies ethnography draws on several established variations of ethnography. Many of these approaches, which are increasingly recognised and applied in LIS research (Hartel et al., 2025), expand the boundaries of multispecies ethnographic inquiry by emphasising embodied, sensory, and relational knowing (Hamilton & Taylor, 2017). Common approaches include:

- Visual ethnography, such as video-making to observe species contact zones (Lynch & Thomas, 2024), or photo diaries and photovoice to explore human–wildlife conflict and biopolitical discourse (Margulies, 2019).
- Sensory ethnography also plays a key role. For example, smell walks have been used to understand how other species experience their environments through sniffing, rolling, and play (Borthwick, 2006). Haptic viewing, which involves using digital surveillance technologies to observe animals remotely, allows human viewers to form emotional and sensory connections with individual animals, enabling forms of digital intimacy while raising ethical questions about representation (Kamphof, 2013).
- Arts-based practices are another important strand. These include participatory theatre, such as the 'Parliament of the species,' which invites multispecies perspectives into urban planning (Olsen, 2022), or animal biographies, which seek to recognise individual animal personhood and challenge human-centred narratives in research and storytelling (Stallwood, 2018; You, 2023).

Combining Ethology and Ethnography

As you begin working in multispecies ethnography, you will quickly realise that studying and describing nonhuman subjects as social agents is not easy. Simply extending human-centred research methods, such as walking interviews and participant observation, to animals risks producing imaginative or overly interpretive descriptions that may not reflect the animals' actual experiences. In multispecies information science, ethnography should not be applied in ways that project human perceptions and experiences about information onto animals. This tendency reflects anthropocentrism, where animals are interpreted solely in relation to human experience (Madden, 2014). To overcome this, some ethnographers combine objective methods developed in natural sciences with ethnographic sensitivity (Lestel et al., 2006).

While ethnography is often regarded as an interpretivist qualitative method, it also possesses a realist dimension that incorporates objective, quantitative measures (Sandstrom & Sandstrom, 1995). As a social scientist, when you begin studying animals and their behaviour in Information Science, you will inevitably find yourself engaging with literature and practices outside the humanities and social sciences, extending towards the natural sciences. One field that may attract your attention is ethology. It is a scientific study of species-specific behaviours using objective, systematic methods (Immelmann, 1980).

Ethologists use a variety of tools and techniques (e.g., video recordings, sound spectrograms, tracking devices, and event recorders) to track animals' movement, vocalisation, and posture. They also carefully document other reciprocal and communication aspects of behaviour between species, including interactions with humans. These methods are used both in natural settings and controlled environments (Immelmann, 1980). The data are often broken down into discrete, classifiable units to identify trends or behavioural repertoires, resulting in tools like the ethogram (detailed behavioural diagrams). Researchers use ethograms to map species' known behaviours, vocalisations, and their umwelten (unique sensory worlds). Ethology thus offers promising avenues for enriching information behaviour research, particularly in embodied and multispecies contexts, where information is exchanged, perceived, and acted upon across species boundaries.

While these methods offer objectivity and precision, interpretation is not entirely free from subjectivity, particularly when human researchers attempt to understand the meaning or intent behind animal actions. Moreover, such studies tend to privilege observable, measurable behaviours while potentially overlooking subjective or relational aspects of animal experience (Hamilton & Taylor, 2017; Van Dooren et al., 2016). Therefore, the extent to which the subjective experiences of animals can be captured using objective scientific criteria remains a contested issue.

This is where ethologically informed ethnography provides a meaningful intervention. It enables researchers to study animals not simply as biological entities, but as social, cultural, and interpretive beings with agency in relation to others.

A key example of this approach is found in the work of John Hartigan Jr (2021), who blends ethnographic methods with ethological techniques to study horse behaviour. Hartigan trained with ethologists to learn structured behavioural sampling (e.g., focal-animal, instantaneous, and continuous sampling) and to use an equine ethogram that included behaviours. This provided a rigorous, data-driven understanding of horse sociality. But Hartigan went further; he interpreted these behaviours not just as fixed patterns but as expressions within ongoing social dynamics, positioning horses as performative, interpretive subjects negotiating their status within the group.

While we can never fully access the inner experience of another species, we can observe how animals respond to their surroundings. We can learn a great deal through interdisciplinary collaboration with scholars in biology, ethology, and animal behaviour studies. Combining their scientific methods with our interpretive approaches creates a better understanding. Also, by spending time with them, we can begin to attune to their ways of being. These practices allow us to make reasonable and ethically grounded inferences about their social, cultural, and informational lives.

Animal Autobiography and Animal Writing

Once you have attuned yourself to animals and their informational worlds, you must then critically consider how to represent their voice and experiences in your research outcomes. Capturing animals' experiences demands both creativity and careful representation. One method is using narrative techniques, such as autobiographical accounts written by ethologists, which rely on thick description and storytelling to communicate animal affect (Lorimer et al., 2019). Ethologists, veterinarians, fiction writers, and scholars across various disciplines contribute to documenting and conveying different aspects of animal life and experiences. Through written works, films, and documentaries, they represent animals. Beyond these forms, many scholars also seek to evoke animal experiences in new ways through performative and creative practices. In this section, I focus particularly on writing combined with other techniques as a means of representing the rich, fleeting, and often non-verbal experiences of other species.

Life writing (e.g., biography and autobiography) has traditionally focused on the individual, human subject, someone who is seen as self-aware, rational, and able to tell their own story, and their life is considered worthy of narrative because it fits into a certain social and political framework. Posthuman ideas challenge this anthropocentric focus, shifting attention to the shared, impersonal life force across species by questioning the dominance of textuality and sight in narrative forms (Braidotti, 2008).

Scholars like Derrida (1999) and Wolfe (2010) deconstruct the assumption that only humans possess narrative selfhood. Their concepts of subjectivity and the *autobiographical animal* open space for recognising nonhuman agency, vulnerability, and communicative forms beyond language, such as scent, gesture, or environmental inscription. Thus, an autobiography is more than just a life story; it is the act of a living being declaring that they are alive and asserting their own existence. This act of self-declaration is something that all living creatures, human and nonhuman, share (Wright, 2021).

So many human traits, like empathy, communication, social structures, and even toolmaking, are found in other animals. However, we have not studied the various language systems of nonhuman animals enough to fully understand their intelligence and inner worlds. While we assume some animals may possess greater capacity than others, our understanding comes from selective studies revealing their social intelligence, conscious choice-making, and ability to form bonds (Cudmore & Hanrahan, 2025).

Within this context, animal autobiography emerges as a speculative narrative mode in which nonhuman subjects are imagined recounting their lives from within complex interspecies lifeworlds (Herman, 2016). Scholarly analysis typically distinguishes between two broad approaches within this genre. The first is allegorical, where the animal narrator functions primarily as a symbolic instrument for critiquing human institutions, values, and social conditions, rather than representing authentic nonhuman subjectivity (Dwyer, 2015). The second is biocentric, an approach that seeks to foreground animal interiority and perceptual difference. These narratives often embrace a hybrid ontological status, challenging readers to engage with the animal as a not-unlike-human co-presence (Savvides, 2013).

Although animal autobiography supports justice for nonhuman animals and echoes Derrida's concerns about their subjugation (Wright, 2021), it still faces critique. The common practice of humans speaking for animals is seen as ethically problematic and reinforces human centrality (Huff, 2017).

In contrast to the representational challenges of animal autobiography, animal writing emerges as a broader methodological practice and critical approach for engaging with nonhuman animals. The animal reorientation critiques the dominance of textuality and vocalcentrism in life writing. So multispecies research heavily criticises the common practice of human authors ventriloquising an animal's voice. This act is seen as an unethical, anthropomorphising fraud that fails to grant true agency to the nonhuman, ultimately forcing the animal into a human narrative (Huff, 2017).

Animal writing uses affective attunement and scientific knowledge to document nonhuman behaviour, perceptual logics, and ecological interactions without reducing them to human meaning systems (Baker, 2019; Chang, 2018;

Piskorski, 2021). In this approach, the researcher describes the environmental context and the interactions within and across species and attends to the senses and perceptions of all species.

Methodologically, animal writing can be refined and integrated with other approaches. In my research with multispecies families, I combined animal writing with the excerpt commentary unit (Emerson et al., 2011) to systematically document their information experiences. To present my findings, I used interactive and multimodal formats, such as hyperlinked narratives and embedded videos and images in text. They enabled readers not only to read but also to see, hear, and sense the multispecies experiences (Solhjoo, 2024a).

Truly animal writing requires moving beyond text. It calls for experimental, multisensory forms, challenging writing and sight as the primary technologies of narrative (Huff, 2017). In my study, I created an art exhibition representing the information experiences of multispecies families (Solhjoo, 2023). Each of the installations featured a central analytical insight of the researcher, surrounded by images and voices of participants. Visitors were invited to contribute their own interpretations via sticky notes, adding a participatory layer. Two displays also included 10-minute video loops illustrating how information flows between humans and animals. The outcome of the research was not a written text but an immersive installation, enabling participants to experience the interwoven lives of humans, animals, and information.

This reorientation does not simply add animals to human stories but fundamentally rethinks the very notion of life being written about. Rich narratives with multispecies installations or immersive, environmental storytelling serve as ethical and representational strategies that resist reinforcing human centrality in information research and practice.

Conclusion

Multispecies methodologies, as introduced in this chapter, encourage the application of holistic, multi-method, context-dependent, and critically reflective perspectives that can be applied across library services, archives, community studies, information studies, information systems, and technology design. Tuning into animals does not neglect human communities. Instead, it recognises responsibilities and new opportunities. It invites you to position your research within the natural, cultural, and social assemblages that you and your community, or the concept under study, are already part of. By creatively engaging with the multispecies turn and its evolving research methods, you can deepen your understanding of information and its underlying theories and practices and help shape meaningful interdisciplinary collaborations.

To see how multispecies research approaches are being practically and conceptually applied, you are invited to explore Section 2 of this volume. It features case studies and conceptual papers where researchers engage with multiple species in their practices and research.

References

Baker, T. C. (2019). *Writing Animals: Language, Suffering, and Animality in Twenty-First-Century Fiction*. Springer International Publishing.

Bastian, M., Jones, O., Moore, N., & Roe, E. (2017). *Participatory Research in More-than-Human Worlds*. Routledge.

Bates, M. (2026). Researching information behaviour among multiple species. In N. Solhjoo (Ed.), *Multispecies Information Science*. Routledge.

Bates, M. J. (1999). The invisible substrate of information science. *Journal of the American Society for Information Science*, 50(12), 1043–1050. https://doi.org/10.1002/(SICI)1097-4571(1999)50:12<1043::AID-ASI1>3.0.CO;2-X

Bernard, H. R. (2018). *Research methods in anthropology: Qualitative and quantitative approaches* (6th ed.). Rowman & Littlefield.

Bonelli C., & Walford A. (2021). *Environmental Alterities*. Mattering Press. http://doi.org/10.28938/9781912729142

Borthwick, F. (2006). Noisy, smelly, dirty dogs: A sensorial autoethnography of living with dogs. In *ACSPRI Social Science Methodology Conference* (1–8), Sydney. Australian Consortium for Social and Political Research Inc. (ACSPRI).

Braidotti, R. (2008). The politics of life as bios/zoe. In A. Smelik & N. Lykke (Eds.), *Bits of Life: Feminism at the Intersections of Media, Bioscience, and Technology* (pp. 179–196). University of Washington Press.

Bruce, C., Davis, K., Hughes, H., Partridge, H., & Stoodley, I. (2014), *Information Experience: Approaches to Theory and Practice* (Library and Information Science, Vol. 9), Emerald Group Publishing Limited. https://doi.org/10.1108/S1876-056220140000010016

Budd, J. M. (2005). Phenomenology and information studies. *Journal of Documentation*, 61(1), 44–59.

Calzati, S. (2025). An ecosystemic view on information, data, and knowledge: Insights on agential AI and relational ethics. *AI and Ethics*, 5, 3763–3776. https://doi.org/10.1007/s43681-025-00665-0

Chang, C. J. (2018). Animal writing: Taiwan's *Dongwu Shuxie*. In K. Tu & T. Russell (Eds.), *Animal Writing in Taiwan Literature* (Vol. 41, pp. 1–18). Taiwan Literature: English Translation Series.

Cohen, S. A., & Cohen, E. (2017). New directions in the sociology of tourism. *Current Issues in Tourism*, 2, 1–20.

Cudmore, K., & Hanrahan, C. (2025). "Humaneness" and the capable animal: A proposal for a non-anthropocentric definition for the subjectivities and integrity of non-human animals. *Human-Animal Interactions*, 13(1). https://doi.org/10.1079/hai.2025.0017

Dashper, K., & Brymer, E. (2019). An ecological-phenomenological perspective on multispecies leisure and the horse-human relationship in events. *Leisure Studies*, 38(3), 394–407.

Day, R. (2026). The animal's importance for neo-documentation: A "Copernican revolution" in understanding information. In N. Solhjoo (Ed.), *Multispecies Information Science* (pp. 53–60). Routledge.

Deleuze, G., & Guattari, F. (1980) 2004. *A Thousand Plateaus. Capitalism and Schizophrenia*. Translated by Brian Massumi. Continuum.

Derrida J. (1999) 2008. *The Animal that Therefore I am*. Translated by M. L. Mallet. Fordham University Press.

Dryzek, J. S. (2016). *The Politics of the Earth: Environmental Discourses*. Oxford university press

Dunn, R. (2021). *A Natural History of the Future: What the Laws of Biology Tell us about the Destiny of the Human Species*. Basic Books.

Dwyer, A. (2015). Animal autobiography and the domestication of human freedom. *Arizona Quarterly*, 71 (2), 1–30. https://doi.org/10.1353/arq.2015.0009

Ekström, B., & Pilerot, O. (2026). Distributed agency and materiality in biodiversity citizen science. In N. Solhjoo (Ed.), *Multispecies information science* (pp. 187–200). Routledge.

Emerson, R. M., Fretz, R. I. & Shaw, L. L. (2011). *Writing ethnographic fieldnotes* (Second edition.). The University of Chicago Press.

Fetterman, D. M. (1998). *Ethnography: Step-by-Step* (2nd ed.). Sage Publications.

Foster, C. (2016). *Being a Beast: Adventures Across the Species Divide*. Metropolitan Books.

Fuller, S., & Lees, F. (2026). The ethics of multispecies information research: an exploratory dialogue. In N. Solhjoo (Ed.), *Multispecies information science*. (pp. 77–90) Routledge.

Given, L. M., Opryshko, A., Julien, H., & Smith, J. (2011). Photovoice: A participatory method for information science. *Proceedings of the American Society for Information Science and Technology*, 48. https://doi.org/10.1002/meet.2011.14504801209

Gorichanaz, T. (2018). Perspective in information behaviour research. In *Proceedings of ISIC, the Information Behaviour Conference*, Krakow, Poland, 9–11 October: Part 1. Information Research, *23*(4), paper isic1803. Retrieved from http://InformationR.net/ir/23-4/isic2018/isic1803.html

Gorichanaz, T. (2020). *Information Experience in Theory and Design*. Emerald Group Publishing.

Gorichanaz, T. (2026). Information experience for the expanding circle: Insights from animal–computer interaction and more-than-human–computer interaction. In N. Solhjoo (Ed.), *Multispecies Information Science* (pp. 40–52). Routledge.

Gorichanaz, T., & Latham, K. F. (2016). Document phenomenology: A framework for holistic analysis. *Journal of Documentation*, 72(6), 1114–1133.

Greyson, D. (2013). Information world mapping: A participatory, visual, elicitation activity for information practice interviews. *Proceedings of the American Society for Information Science and Technology*, 50. https://doi.org/10.1002/meet.14505001104

Hamilton, L., & Taylor, N. (2017). *Ethnography after Humanism: Power, Politics and Method in Multi-Species Research*. Palgrave Macmillan

Haraway, D. (2003). *The Companion Species Manifesto: Dogs, People, and Significant Otherness*. University of Chicago Press.

Haraway, D. (2016). *Staying with the Trouble - Making kin in the Chtulucene*. Duke University Press.

Hartel, J., Lu, A., Montague, K. E., Solhjoo, N., & Stewart-Robertson, O. (2025). Ethnographic variations. *Proceedings of the Association for Information Science and Technology*, 62(1), 1235–1240. http://doi.org/10.1002/pra2.1373

Hartigan Jr, J. (2021). Knowing animals: Multispecies ethnography and the scope of anthropology. *American Anthropologist*, 123(4), 846–860.

Herman, D. (2016). Animal autobiography; or, narration beyond the human. *Humanities*, 5(4), 82.

Hirskyj-Douglas, I., Read, J., & Cassidy, B. (2015). *Doggy Ladder of Participation (DLOP). In Paper Presented at the British Human Computer Interaction*, Lincoln.

Huff, C. (2017). After auto, after bio: Posthumanism and life writing. *a/b: Auto/Biography Studies*, 32(2), 279–282.

Immelmann, K. (1980). *Introduction to Ethology*. Springer.

Jones, A., & Hoskins, T. K. (2016). A mark on paper: The matter of indigenous settler history. In C. Taylor & C. Hughes (Eds.), *Posthuman Research Practices in Education* (pp. 75–92). Palgrave Macmillan.

Kamphof, I. (2013). Linking animal and human places: The potential of webcams for species companionship. *Animal Studies Journal*, 2(1), 82–102.

Käufer, S., & Chemero, A. (2015). *Phenomenology: An Introduction*. Polity Press.

Kim, J., Kim, R., Martins, J., & Karana, E. (2024). Becoming microbes: An approach to cultivating microbial sensibilities in biodesign. In *Proceedings of Design Research Society (DRS 2024)*. Paper 298. https://dl.designresearchsociety.org/drs-conference-papers/drs2024/researchpapers/298/

Kirksey, S. E., & Helmreich, S. (2010). The emergence of multispecies ethnography. *Cultural Anthropology*, 25(4), 545–576.

Latour, B. (2007). *Reassembling the Social: An Introduction to Actor-Network-Theory*. Oxford University Press.

Lestel, D., Brunois, F., & Gaunet, F. (2006). Etho-ethnology and ethno-ethology. *Social Science Information*, 45(2), 155–177.

Lestel, D., Bussolini, J., & Chrulew, M. (2014). The phenomenology of animal life. *Environmental Humanities*, 5(1), 125–148.

Lewis, T. E., & Owen, J. (2020). Posthuman phenomenologies: Performance philosophy, non-human animals, and the landscape. *Qualitative Inquiry*, 26(5), 472–478.

Lorimer, J., Hodgetts, T., & Barua, M. (2019). Animals' atmospheres. *Progress in Human Geography*, 43(1), 26–45.

Lupton, D. (2019). Toward a more-than-human analysis of digital health: Inspirations from feminist new materialism. *Qualitative Health Research*, 29(14), 1998–2009.

Lynch, J., & Thomas, H. (2024). Digital Technology in outdoor and environmental education: Affects, assemblages and curriculum-making. *Australian Journal of Environmental Education*, 40(2), 288–304.

Madden, R. (2014). Animals and the limits of ethnography. *Anthrozoös*, 27 (2), 279–293.

Mancini, C., & Lehtonen, J. (2018, June). *The emerging nature of participation in multispecies interaction design. In Proceedings of the 2018 Designing Interactive Systems Conference* (pp. 907–918). Hong Kong, China, 9–13 June 2018. https://doi.org/10.1145/3196709.3196785

Margulies, J. D. (2019). On coming into animal presence with photovoice. *Environment and Planning E: Nature and Space*, 2(4), 850–873. https://doi.org/10.1177/2514848619853060

Nagel, T. (1974). Philosophical review what is it like to be a bat? *The Philosophical Review*, 83(4), 435–450. http://www.jstor.org/stable/2183914

Nimmo, R. (2011). Actor-network theory and methodology: Social research in a more-than-human world. *Methodological Innovations Online*, 6(3), 108–119.

Noorani, T., & Brigstocke, J. (2018). More-than-human participatory research. In K. Facer & K. Dunleavy (Eds.), *Connected Communities Foundation Series* (pp. 1–40). University of Bristol/AHRC Connected Communities Programme.

Ogden, L. A., Hall, B., & Tanita, K. (2013). Animals, plants, people, and things: A review of multispecies ethnography. *Environment and Society*, 4(1), 5–24.

Olsen, S. (2022). Co-creation beyond humans: The arts of multispecies placemaking. *Urban Planning*, 7(3), 200–211. https://doi.org/10.17645/up.v7i3.5288

Piskorski, R. (2021). Introduction: Animal (As) writing, writing (As) animal. *Word and Text, a Journal of Literary Studies and Linguistics*, 11, 5–12.

Pollak, A. (2017). Visual research in LIS: Complementary and alternative methods. *Library & Information Science Research*, 39(2), 98–106.

San Martín, J., & Peñaranda, M. L. P. (2001). Animal life and phenomenology. *The Reach of Reflection: Issues for Phenomenology's Second Century*, 2, 342–363.

Sandstrom, A. R., & Sandstrom, P. E. (1995). The use and misuse of anthropological methods in library and information science. *The Library Quarterly*, 65(2), 161–199.

Savolainen, R. (2019). Elaborating the sensory and cognitive-affective aspects of information experience. *Journal of Librarianship and Information Science*, *52*(3), 671–684. https://doi.org/10.1177/0961000619871595

Savvides, N. (2013). Speaking for dogs: The role of dog biographies in improving Canine Welfare in Bangkok, Thailand. In M. DeMello (Ed.), *Speaking for Animals: Animal Autobiographical Writing* (pp. 231–243). Routledge.

Schreiber, T. (2017). E-learning objects and actor-networks as configuring information literacy teaching. *Information Research*, 22(1), CoLIS paper 1605. Retrieved from http://InformationR.net/ir/22-1/colis/colis1605.html

Shores, C. (2017). What Is It Like To Become a Rat?: Animal Phenomenology through Uexküll and Deleuze & Guattari. *Studia Phaenomenologica*, 17, 201–221. https://doi.org/10.5840/studphaen20171710

Smith, J. A., & Mitchell, R. W. (2012). *Experiencing Animal Minds: An Anthology of Animal-Human Encounters*. Columbia University Press.

Solhjoo, N. (2023). The photo exhibition: A visual exploration of information in our shared lives with companion animals. *Information Matters*, 2(4). https://doi.org/10.2139/ssrn.4450402

Solhjoo, N. (2024a). Information Experience Within Multispecies Families: Understanding Shared Life with Companion Cats and Dogs (doctoral dissertation). Open Access Te Herenga Waka–Victoria University of Wellington. https://openaccess.wgtn.ac.nz/articles/journal_contribution/The_In-between_Information_Experience_within_Human-Companion_Animal_Living/24986298

Solhjoo, N. (2024b). Knowing within multispecies families: An information experience study. *Journal of Information Science*. https://doi.org/10.1177/01655515241268845

Solhjoo, N., Goulding, A., & Krtalic, M. (2022). Companion animal representation and voice in information behavior research. *Library & Information Science Research*, 44 (4). https://doi.org/10.1016/j.lisr.2022.101198

Solhjoo, N., & Hartel, J. (2026). When animals teach: Lessons for information pedagogy. In N. Solhjoo (Ed.), *Multispecies Information Science* (pp. 61–76). Routledge.

Spradley, J. P., & McCurdy, D. W. (1972). *The Cultural Experience: Ethnography in Complex Society*. Science Research Associates.

Stallwood, K. (2018). Topsy: The elephant we must never forget. In A. Krebber & M. Roscher (Eds.), *Animal Biography*. Palgrave Studies in Animals and Literature. Palgrave Macmillan. https://doi.org/10.1007/978-3-319-98288-5_12

Suorsa, A. (2024). Embodied and dialogical basis for understanding humans with information: A sustainable view. *Journal of the Association for Information Science and Technology*, 75(13), 1466–1479.

Suorsa, A. R. (2015). Knowledge creation and play–a phenomenological approach. *Journal of Documentation*, 71(3), 503–525.

Tarcan, B., Pettersen, I. N., & Edwards, F. (2022). Making-with the environment through more-than-human design. In D. Lockton, S. Lenzi, P. Hekkert, A. Oak, J. Sádaba, & P. Lloyd (Eds.), *DRS2022: Bilbao*, 25 June–3 July, Bilbao. https://doi.org/10.21606/drs.2022.347

Uexküll, J. Von. (1934) 2010. *A Foray Into the Worlds of Animals and Humans, with a Theory of Meaning*. Translated by J. D. O'Neil. University of Minnesota Press.

University of Brighton, Centre for Arts and Wellbeing. (2024, June 3). *Abstracts: Exploring human-Animal and Multispecies Relations – Risk Taking in Research Methods Symposium*. https://blogs.brighton.ac.uk/artswellbeing/2024/06/03/abstracts-exploring-human-animal-and-multispecies-relations-risk-taking-in-research-methods-symposium/

van der Linden, D. (2022). *Animal-centered design needs dignity: A critical essay on ACI's core concept. Paper Presented at the Proceedings of the Ninth International Conference on Animal-Computer Interaction*. https://doi.org/10.1145/3565995.3566028

Van Dooren, T., Kirksey, E., & Münster, U. (2016). Multispecies studies: Cultivating arts of attentiveness. *Environmental Humanities*, 8(1). https://doi.org/10.1215/22011919-3527695

VanScoy, A., & Evenstad, S. B. (2015). Interpretative phenomenological analysis for LIS research. *Journal of Documentation*, 71(2), 338–357. https://doi.org/10.1108/JD-09-2013-0118

Vincent, A. (2019). Dog parks as an institutional resource for social capital in the urban neighborhood. *Society & Animals*, 29(5–6), 1568–5306.

Westerlaken, M., & Gualeni, S. (2016). Becoming with: towards the inclusion of animals as participants in design processes. In *Proceedings of the Third International*

Conference on Animal-Computer Interaction (pp. 1–10). https://doi.org/10.1145/2995257.2995392
Wolfe, Cary (2010). *What Is Posthumanism?* University of Minnesota Press.
Wright, S. A. (2021). Jacques Derrida and the autobiographical (Non-human) animal: An analysis of (false) animal autobiographies. In K. Maiti (Ed.), *Posthumanist Perspectives on Literary and Cultural Animals* (pp. 87–98). Springer International Publishing.
You, C. (2023). Genre, tradition and renewal: Animal autobiography and poetics of the multicentric self. *Humanities and Social Sciences Communications*, 10(1), 1–8.
Yu, L., & Liu, Y. (2022). Information experience as an object of LIS research: a definition based on concept analysis. *Journal of Documentation*. https://doi.org/10.1108/JD-10-2021-0195

SECTION 2
Multispecies Studies

Niloofar Solhjoo

Building on the theoretical, ethical, and methodological foundations of Multispecies Information Science established in Section 1: *Fundamentals*, the second section brings these ideas to life through a diverse and creative array of case studies and conceptual research. Together, these creative studies deepen and expand your understanding of what it means to engage with information in a more-than-human world.

They share stories about everyday life and professional settings, from moments of joy and leisure to experiences of crisis and vulnerability, from the deep past to speculative futures shaped by our entanglements with animals. Across these varied stories, human and nonhuman lives intersect in meaningful ways, revealing how information appears, flows, and acts through multispecies environments, relationships, and practices.

The influence of animals on my academic life and work was shaped by my experience as an animal health information professional in a veterinary clinic in Tehran, Iran, in 2015, where I developed information literacy resources to help clients better understand their animals' health and critically evaluate the information they used in caregiving. My close relationships with the community involved in animal healthcare, along with my personal bonds with animals—particularly during the emotionally difficult loss of a beloved companion dog—gradually nurtured the idea of multispecies information science in my research. This ultimately led to my doctoral research at Te Herenga Waka—Victoria University of Wellington, New Zealand, titled *Information Experience within Multispecies Families*, which explored understanding in shared life with companion cats and dogs (Solhjoo, 2024).

DOI: 10.4324/9781003583424-9

The research was neither solely a study of human information behaviour nor merely of animal information behaviour; rather, it was focused on multi-species information experience (Solhjoo, 2024). It concludes with three core lessons: Loving, living, and learning. *Loving* refers to the affective forms of information within multispecies families and how mutual care becomes a primary mode of understanding each other. *Living* is information embedded in everyday bodies, objects, and the shared physical space of the home, showing that becoming informed within a multispecies family requires integrating information processes across living and non-living entities. Finally, *learning* describes cognitive processes of participatory sense-making in human–animal play, routines, and life experiences.

All the chapter authors in this section have similarly empathetic stories about the animals in their lives. They have lived, loved, and learnt with other species, carrying these experiences into LIS and their academic work.

Dirk van der Linden opens this section by exploring humour and co-existence in *Live, Laugh, Love Urban Wildlife*, analysing how laughter during human–wildlife encounters in the UK can inform pathways to equitable multispecies cities. Bhuva Narayan then immerses us in *Urban Dog Parks as Information Grounds*, using sensory ethnography to argue that Sydney's dog parks function as vital nodes in a city-wide network where humans and dogs co-create community and information.

The theme of information behaviour studies continues with Monika Krakowska and Magdalena Zych's *Frosted Muzzles and Digital Bonds*, which examines the affective information practices and interspecies testimonial injustices that arose during a crisis-driven animal adoption event in Poland. Yazdan Mansourian offers a quiet contrast in *'Building a Bonsai Bond,'* reflecting on the pursuit of perfection and the deeply embodied information experience within a person–plant partnership. Björn Ekström and Ola Pilerot then scale up in *Distributed Agency and Materiality in Biodiversity Citizen Science*, using a posthumanist frame to show how information practices in monitoring Swedish species are co-produced by a web of human, organic, and technological actors.

Isto Huvila looks to the past to inform the present in *Habitats of Archaeological Knowledge*, using archaeology as a key example to derive strategies for studying any multispecies information habitat. Finally, Erik Radio concludes this section by gazing outward in *Massive Unseen Companions and the Documentary Project*, using the metaphor of a binary star system to challenge our classificatory practices and explore how entities are known only through their relations to others.

Together, Section 2: *Studies*, provides a compelling and compassionate roadmap for engaging with the multi-species world in information research and practice.

Reference

Solhjoo, N. (2024). Information experience within multispecies families: Understanding shared life with companion cats and dogs (doctoral dissertation). Open Access Te Herenga Waka–Victoria University of Wellington. https://openaccess.wgtn.ac.nz/articles/journal_contribution/The_In-between_Information_Experience_within_Human-Companion_Animal_Living/24986298

8

LIVE, LAUGH, LOVE URBAN WILDLIFE

Dirk van der Linden

Introduction

There is surprisingly little research on the role of laughter in human–animal relations, even with the ability to laugh preceding the ability to speak and evidence increasingly showing nonhuman animals laugh, too (Panksepp, 2005; Winkler & Bryant, 2021; Davila-Ross & Palagi, 2022). Of the few researchers who braved this topic, Valeri (2006) asked whether laughter can serve as 'a form of interspecies communication, to strengthen social bonds and express positive affect between members of different species,' and Mitchel & Sinkhorn (2014) posited that laughter with or at companion animals is very much grounded in our being social partners, in line with understandings of laughter serving as a behavioural mechanism for social bonding (Dunbar, 2022); yet, even since then, little research has sought to explore to what extent laughter occurs in encounters with nonhuman species.

Laughter, though, is such a primal and important affair. As Carus (1898, p. 251) wrote,

> Life is serious, and if we could see all the misery of life at once it would so oppress us that we would long to die. But because life is serious, and because we need a buoyant spirit to fight the struggle of life bravely, we need as a temporary relief from time to time a hearty laugh.

Understanding when and where we laugh, and how that affects the way we coexist, ought thus to be a worthwhile subject of study in itself. Laughter, it has been argued, ought to be seen as a language in its own right (Ginzburg et al., 2020), which makes conceptualising it as a language crossing the boundaries

DOI: 10.4324/9781003583424-10

between species even more interesting. When we bring in the potential for laughter to strengthen interspecies bonds, it seems entirely justified to explore how we laugh with and at nonhuman animals we share our lives with. From a multispecies information science point of view, those animals that share 'our' cities with us but very much live their own lives—urban wildlife—should be of particular interest. This covers a variety of nonhuman life in a variety of environments: From our homes and gardens to public green and grey spaces alike. Similarly, urban wildlife covers a broad variety of animals, from those widely loved to those almost universally despised (cf. Perry et al., 2020).

For this specific investigation, we intend not only to study an interesting phenomenon but also to show how adopting the notion of a multispecies information landscape, where the focus moves beyond just the human, in line with Solhjoo (2026, Chapter 7), allows for research to be guided to move away from an anthropocentric nature. Specifically, we will be focusing on a specific physical layer (the shared urban environment) and a specific behaviour of significance (laughter in multispecies encounters) and, using reflexive thematic analysis, construct a theory of how this shared multispecies urban umwelt may be far more similar for humans and nonhumans than thought.

This focus links into the study of information behaviour (as in, 'the [many] ways in which individuals interact with information, in particular, the ways in which they experience and utilise information' (Bates, 2010)). While early research in everyday life information behaviour tended to focus on the more negative (illness, unemployment, and other things disrupting us from our lives), there are many more behaviours to study to paint a full picture of just how information guides and perhaps shapes our very behaviours. That scope of information behaviour research has broadened to include what Kari and Hartel (2007) call the 'higher things in life': Pleasurable or profound experiences and phenomena that go beyond just the routines of daily survival to how people share happy information in casual settings as a way to connect, test, or strengthen social relationships (Tinto & Ruthven, 2016).

To that end, this chapter provides a case study exploring how and where encounters with urban wildlife have led to laughter, analyses what seem to be the shared characteristics of such encounters, and posits how we might think of laughter in human-urban wildlife encounters as a premise for increased understanding and coexistence.

Method

Based on an existing dataset of urban wildlife encounters across the UK (van der Linden et al., 2025), we wanted to look at those encounters where people recalled laughing to some degree. Having people describe particularly memorable encounters with nonhuman life provides us with access to more than just information on the 'surface'; it provides us with a way to phenomenologically interpret these encounters and go into the 'information below,' as highlighted by Gorichanaz (2026, Chapter 3) of this volume. Our inclusion criteria were simple: Encounters where people unambiguously laughed. We searched for encounters (positive and negative alike) where some variation of the term 'laugh*' was used.

This resulted in a set of 130 encounters (out of a total of 2,010 positive encounters) for analysis, covering encounters with squirrels, foxes, gulls, pigeons, hedgehogs, rats, ducks, deer, badgers, a smattering variety of birds, and also one cow and a horse. We removed one encounter that talked about roadkill, as upon closer inspection, it talked not of laughter at the dead (thankfully) but expressed disgust at the cruelty of seeing 'humans' laughing at driving over an innocent herd of ducklings. Removing that left us with 129 encounters to analyse.

We then conducted a reflexive thematic analysis (Braun & Clarke, 2019) on this subset of encounters, focused on creating an understanding of what makes people laugh or react with humour to an encounter with urban wildlife. As a multispecies information science 'practitioner,' we then very much took a phenomenologist approach to understanding these encounters and the animal life therein (as detailed in Solhjoo (2026, Chapter 7)), looking at the phenomenon of two living beings interacting (indirectly, through the recalled encounter), and using reflexive thematic analysis to attempt to describe just *how* those encounters are experienced by stepping into a shared multispecies perspective through constructing a theory of how human and nonhuman urban wildlife live together and affect one another.

Findings

We constructed the following themes: Laughter from observing animal antics, essentially the nonhuman animals going about their lives in our human world; irreverent interactions, nonhuman animals going about their lives very much interacting with our human world; and reflective responses, where finally animals going about their lives make us reflect and reposition our own position in life.

Theme: Animal Antics; or, Laughing from How Animals Live

Seeing, or watching, animals do their thing leads to laughter in a variety of encounters. Many of these encounters are simply amusement or enjoyment from the actual, sometimes unexpected, act of seeing wildlife. We see wildlife in a variety of urban settings, whether around our own homes and gardens, like some people noting they *have wood pigeons that come to our garden and they make me laugh because they are quite clumsy. I love watching them. They are so pretty and make me feel happy*, or others who saw *a squirrel bury a nut in the plant pot and then a few months later, come back and scavenge for it. Made me feel at one with nature and laugh at how they were scrambling for the nut.*

Similarly, as we are out and about in the city or moving from one place to another, we encounter wildlife that sparks laughter, covering a wide variety of different wildlife—even animals that are typically despised, like one person noting they

> saw two little rats on the underground, one more in London, around 5:30 am. I didn't interact [with them] but I remember laughing and them sparking some joy for me as they ran around playing together whilst London was still asleep.

Or another person noting they saw *pigeons arguing over some food someone had dropped. I just watched it (with my toddler, who thought they were brilliant!) and it made me laugh the way they were arguing with each other*. Obviously, such casual encounters exhibit an expected degree of anthropomorphism (most people are not trained ethologists, after all), which likely facilitates being able to laugh at these encounters by conceptualising them from a human point of view. The wildlife we perceive provides us with information, but we can only interpret it through the lenses that we have. Shared encounters with family and loved ones, especially younger humans, who perhaps have less of the bias and weariness many of us go through life with, also seem to spark laughter, such as a parent noting they saw

> seagulls on the school run passing a boating lake, my 2 year old loves the noises and way they move. We feed them seed and he laughs like crazy when they start squabbling (in a nice way). It was brilliant to see him so happy with something so trivial to me.

Unlike those encounters where an anthropomorphic interpretation of the animal behaviour aids in allowing us to laugh, here the joy of seeing others *enjoy* (perhaps for the first time) the nonhuman world sparks joy in ourselves.

Seeing animals at an unexpected moment can make us laugh, like one person noting they saw *a hedgehog crossing the road while I was on a late night walk. Made me laugh with his little waddle*, and even opted to share this moment as they *made a video and sent [it] to friends*, or another person who noted they

> saw two ducks (male and female) on my walk home from the city. They were sat on a grassy area by the side of a busy road and it looked really out of place. I stopped and took my headphones out as though I was going to talk to them which made me laugh because it was a strange interaction.

Unexpected animals similarly can stimulate laughter (or, more formally, as Warren and McGraw (2016) argue, when benign violations of expectations occur that are still 'okay'), like a person who noted they *once saw a badger in my university halls after a night out. It made me laugh as I'd never seen a Badger before.*

While the exact reason for why these encounters made people laugh may vary, from seeing an unexpected animal to seeing them in an unexpected place or simply sharing in the joy and excitement of other humans experiencing it, a common thread seems to be that we laugh at these encounters where animals clearly live *along* us in the human world, very much going about their own affairs.

Theme 2: Irreverent Interactions; or, Laughing at How Animals Live but Are Not Beholden to the Human World

Whereas before we laughed at seeing animals live *along* us, there seems to be a meaningful difference in seeing them truly *among* us, interacting in some way with the human world, and in doing so, often showing just how little they care for the social expectations we set out for how to behave with one another.

This can be as simple as having to laugh from seeing an animal make a 'human owned' space its own and not caring, whether in private urban spaces, like one participant noting how

> there was a fox just brazenly having a snooze under a tree in our front garden in the middle of the day, my husband and I were standing over him right up close, watching amazed, and he was not bothered by us at all. It made us laugh as he didn't care [that] we were there, and it was as if he owned the place.

Or in public urban spaces, like another participant noting how

> [three] ducks were walking down the street where we live. They didn't fly off when they saw people. Me and my family went into our front garden just to watch them walk past. It was funny to see ducks just walking down a street. It made us all laugh and we felt happy to see it.

Some animals, from the human perspective, have very little regard for how humans typically feel 'we' should all behave in the human-built and 'owned' urban environment and still make us laugh, even if that humour would be considered juvenile by some, such as a person noting that *I have seen horses in my town that have taken a massive poo in the town centre which made me laugh.* Indeed, it is that kind of 'violation' of what we would typically expect to happen or deem acceptable (Warren and McGraw, 2016), while not going 'too far,' that makes these situations humorous to us. At the same time, sometimes we laugh also because perhaps we realise that these urban spaces are not ours at all, as one person noted who

> was chased once by highland cows, I was literally terrified for my life. I had my dog with me and they had babies. They stopped running after me when I went down a bank, and I was so relieved I burst out laughing!

Beyond laughing at animals living among us and perceiving them not to care about us, what we believe our spaces are for, and how they ought to be used, we laugh and perhaps even admire the irreverence of wildlife living among us. For example, as one participant noted

> I was in B&Q car park eating a Burger King Whopper and fries with the window down slightly because it was hot. Immediately a seagull appeared about 1.5-2 metres away from my car. I could feel its eyes on me whilst I ate but every time I turned to look at the bird, it would turn its head and pretend it hadn't been staring at me. It made me laugh and I gave it some fries.

Even if their behaviour is not necessarily intentional, we still laugh, as one person noted they were once '*face-slapped by a low flying gull flipper next to the harbour. It was just flying off a railing, not scared or attacking. It [made] me and my son laugh a lot.*'

Especially when this kind of irreverent behaviour happens to others, we tend to laugh, in a display of *schadenfreude*, like one participant noting how they saw *my friend tried to feed a fox a burger it ignored the burger and stole the chocolate from her other hand it was very funny and made me laugh*, and in some cases even go beyond mere laughter to admiration, like another person who recalled when a *herring gull swooped down and took a sausage roll from a man's hand. It made me laugh to see this bird being brave and resourceful*. Again, some degree of anthropomorphism (in this case, positioning the gull as brave and resourceful), together with violations of what we expect (should) happen around us, aids heavily in conceptualising these encounters as humorous.

It thus seems that one key aspect where urban wildlife makes us laugh is from these encounters where they very much live *among* us as *part* of the human world, rather than merely alongside it, and how our laughter at the behaviour they express seems to be perhaps rooted in how we realise they are not beholden to the social contract we all struggle under and wish we could forget every now and then.

Theme 3: Reflective Responses; or, Laughing and Then Thinking

Some encounters make us laugh but push us beyond mere laughter or enjoyment. We see the successes or challenges of the wildlife sharing our cities with us, and, through laughter, think and reflect on just how alike we are. This ranges from initial laughter to reflection and sympathy for animals' plight, such as one person noting who

> had a gull stuck in my backyard all day /…/ could hear it crying all day, it was a baby gull, so probably for his mother and it made me feel extremely sympathetic to its pain and abandonment. It made me laugh for a second but I was mostly saddened. It kind of changed my perspective on gulls again, since they are mostly seen as worse than rats, when they're just animals and part of life too.

While we might initially laugh at such encounters, possibly at just the unexpected sounds, when we understand what is happening, when we move from the mere 'data' of a gull's sounds to interpreting it as the information of a gull crying for their mother—even if again needing to anthropomorphise to relate by thinking how *we* would feel in such a situation—we stop laughing and move on to deeper emotions and cognitive actions.

Such reflective thought, where we realise the arbitrary 'us' and 'them' distinction holds little meaning, happens in a variety of encounters. Indeed, there are more encounters like the above where someone

> came across a seagull that had a crisp packet stuck on its head. This was pretty funny at first but also dangerous for the bird, so I assisted in getting said crisp packet removed from its head. This made me both laugh and also feel good that I had helped an animal from a potentially dangerous situation.

Animals of all kinds, perhaps, are more like us than we realise, as one person noted when they saw a *swan interact[ing] with its signets, made me laugh as it was very similar to human parent & small child interaction with the swan trying to get the signets in order*.

Humorous encounters also allow us to relativise our human lives, as one person noted when they were

> walking [through] the local park when I saw 3 squirrels chasing each other up and down the trees. It made me laugh as they appeared to be having a fun time. I stopped and watched them for a bit as I was having a stressful day and they helped put things into perspective.

Such encounters are not uncommon, as laughter resulting from either intentionally observing or suddenly coming face to face with wildlife lets us reflect, like another person noting one time in the city

> out of nowhere a pigeon went above me and [pooped] on my head. I was with family and to my surprise everyone started to laugh and thought it was really funny. It made me laugh and feel a little silly in a positive way. To this day we talk about this incident and are amused. The pigeon didn't notice or care what it did and went on its way.

Indeed, this shows how encounters that make us laugh because of behaviour perceived to be irreverent can go further and make us reflect.

But humorous encounters can lead to deeper reflection, blurring the boundaries between human and nonhuman. For example, one person noted how once they were in a park, watching a squirrel who

> *was jumping around with a lot of food. It kept dropping what it was carrying and going back to pick it up even though it had too much. I was standing there watching its antic[s] and It made me laugh. I thought I know exactly how you feel squirrel. Trying to carry to much at a time.*
>
> (*emphasis added*)

The way the animal behaved in this encounter, although funny at first glance, offered a powerful reflective lens to its observer, suddenly becoming closer and realising how we share difficulties in life. Indeed, we see how some concepts we might otherwise consider deeply 'human' are evident in animal lives too, as one person who noted they

> laughed at seeing a rat trying to carry a rubbish packet of a chocolate bar, but the packaging was heavy and it was jumping trying to carry it, it made my day knowing that not only humans struggle in life.

As before, people clearly interpret these encounters from a human perspective. But we would argue that it is not necessarily bad, nor is it evidence per se of an anthropocentric attitude. Indeed, as Taylor (2011) argued, anthropomorphism is inevitable as we interpret the world around us through our own embodied materiality, but that in no way has to lead to (or be based on) assumptions of human superiority.

The encounters we have discussed so far, indeed, seem to bring us closer to the nonhuman and help us downplay the notion of human superiority, as what we thus see is that wildlife encounters that make us laugh can help us relativise our own human lives but also realise just how alike 'we' and 'they' are, and perhaps even realising that it is more us who are like them than they are like us.

Theme 4: Bi-directionality; or, What If We're the Butt of Their Jokes?

While we only asked *people* to recall encounters that made them laugh (mostly as we would not profess to being fluent in any kind of animal language), it would be silly not to consider whether these encounters have the potential to make animals laugh at us. If we consider that it might equally be wildlife observing us and perhaps deriving something not entirely unlike our experience of humour (as Panksepp (2005) already argued is not without its merit) from those observations, how might we then change our own perspectives? This, for obvious reasons, is rather speculative, but it may be a useful way to rethink and challenge human-focused multispecies encounters and broaden our understanding of the agency of animals in such encounters and what it would take to do so (cf. Haraway, 2008; Despret, 2016).

Doing so links in with the broader shift of the 'animal turn,' which, as Wolfe (2010) argues, demands transdisciplinary efforts and reflexivity from us as researchers. Studying animals, and in this case, information behaviours such as laughter, is not the exclusive domain of any one discipline, but requires all of us (and *Multispecies Information Science*) to engage with each other. In doing so, we might not only gain new insights into animal information behaviours, such as the one we discuss in this chapter, but also critically challenge the assumptions and limitations of our own conceptual and methodological frameworks.

So, what could information interaction look like in a multispecies context? Consider some of the encounters recalled by participants where the animals were not passively observed but were an active presence in the encounter that sparked potentially shared information behaviours. For example, one participant recalled

> /.../ watching tennis at Eastbourne, which is a big international event. A seagull landed on the net and just seemed to watch everyone for a few seconds. When the ball boy shooed it away, it flew straight across the court and stole a sandwich that a spectator had been eating.

The gull's observed behaviour here, and the way it is interpreted after, might be interpreted not just as an animal exhibiting (expected) natural behaviour, but perhaps as an animal deeply interwoven in human-dominated settings that has become aware of the social rules of engagement and intentionally chooses to violate them, in a way that might be interpreted as humorous by both. Who are we to say that the gull does not find the human spectacle and response to its interruption amusing?

Indeed, many other participants recalled encounters where they felt observed by wildlife. As one participant noted,

> whilst having coffee one morning we looked up and staring at us through our patio door was a rather tatty fox. It was very curious and didn't move when we went to the window for a closer look. It made us laugh because it seemed to be as interested in us as we were in him.

Another participant recalled how

> a crow was sat on my windowsill looking in at me. I didn't interact with him except stare at him as I didn't want to scare him away. It made me laugh though as he seemed very interested in what was going on inside.

Such encounters show that people already conceptualise themselves not only as observers but also as the observed in multispecies encounters. There is a perhaps uncanny sensation of the animals watching to see what we do and, in turn, responding to it, very much exemplifying what Haraway (2008) frames as the inevitable 'becoming' within our multispecies relationships.

If we entertain the (speculative) possibility that animals might laugh at us, or at the very least derive something not entirely unlike humour from observing our behaviours and antics, as a field we might help further challenge anthropocentric framings and see ourselves as part of a shared multispecies information landscape of attention, curiosity, and indeed, laughter.

Discussion

This chapter started by asking what and how encounters with wildlife lead to laughter, and what characteristics might be shared in such encounters. Effectively, what *is* the role of laughter in human–wildlife relations? As we summarise in Figure 8.1, laughter happens across a diversity of encounters, in different settings, at different times, with different nonhuman species, recalling the importance of context in understanding laughter (Rychlowska et al., 2022). Critically, we don't tend to laugh *at* animals as much as we laugh *because* of or *with* them. What seems to be a red line across the different recurring themes we constructed is that it is perhaps the *lack* of difference between animals and ourselves that is most salient: We laugh at seeing human animals living alongside us in the human world, seemingly careless to us going about our business; we laugh at their irreverence when they interact with us or the human world; and seeing their struggles and successes makes us realise it is perhaps more *us* that is like *them*, rather than the other way around.

That is describing when and where laughter happens. But what might the *role* of laughter be in all this? We posit that one, if not *the* most important,

FIGURE 8.1 Narrative of our thematic analysis showing how encounters with urban wildlife make us laugh and think, panels based on specific encounters. (Images generated using deep.ai).

aspect of laughter in human–wildlife relations in urban settings is *normalisation*; that is, laughter seems to play some role in the way we (i.e., the human observer) normalise and integrate natural animal behaviour in intrinsically human-dominated environments.

One might be forgiven for thinking they have strayed into a different book and have read a chapter showing a traditional human–animal relations study, with little *information*, though. However, we argue that this thematic analysis shows the importance of multispecies information science, as the role of laughter we have described can happen only when wildlife is there to *provide* us with information to enable those processes. That is, we have now constructed a narrative positioning urban wildlife first and foremost as *information providers*. That is, they provide us, knowingly or unknowingly, with information that allows us to confront the triviality and sometimes absurdity of our own human experiences. In doing so, we grow through these kinds of multispecies encounters—the very cluster of experience that makes up the self (Gorichanaz, 2020) is strengthened through relational experiences, especially when they are with nonhuman entities, like urban wildlife here.

Reconceptualising urban wildlife as such essential information providers means that they are not simply subject to our human observation, nor only creating their own animal narratives, but co-shaping the narrative we apply to *our* lives: A truly multispecies turn (Solhjoo, 2025). As we have argued through this thematic analysis, it is not so much that animals' lives are 'other' to us, but rather that our lived experiences are far more alike to 'them' than we might wish to profess, something we argue might just be key to building that critically needed coexistence with wildlife (cf. Carter and Linnell, 2023). It is exactly that realisation which we need to promote equitable and just multispecies coexistence. Curtin and Kragh (2014) have argued that wildlife tourism has the potential to reawaken a 'deeper ecological sub-consciousness brought about by a society that is disconnected from nature.' Imagine, then, what effects we might achieve if we only need to become tourists in the very places we live rather than engage in otherwise potentially problematic tourism (cf. Kline et al., 2023).

A multispecies information science turn is thus critically needed here, covering at least two aspects of laughter in these encounters: Understanding first how seeing wildlife as information providers triggers laughter that can help us coexist better, but more importantly, understanding if and how *we* can be similar information providers to wildlife, offering them the same benefits we gain from these encounters (through bidirectional exchange of information, as argued before as an important future endeavour in other more-than-human efforts (van der Linden, 2021)). A multispecies turn is necessary here, as research on how human information is parsed by wildlife is very much rooted in fundamental animal behaviour research, for example, understanding differences between threats or lack thereof (Goumas et al., 2022).

Conclusion

What is needed, then? Multispecies information science can help us build towards a more equitable coexistence by providing us with a frame to understand how urban wildlife and nonhuman animals in general generate information that contributes to laughter; and eventually, hopefully, understand how urban wildlife, and again, nonhuman animals in general, can be(come) consumers of such information. Going back to the little extant research on the role of laughter in human–animal relations and the questions it posed, we need to ask ourselves, can laughter strengthen social bonds and express positive affect between 'us' and urban wildlife? Can it help us become social partners, coexisting equally in the city? We say an emphatic yes. *Multispecies Information Science*, perhaps then, can bring a wake-up call that focusing on laughter and subsequent joint happiness in multispecies relations is the way forward for how we coexist. What a time to be alive.

References

Bates, M. J. (2010). Information behavior. *Encyclopedia of Library and Information Sciences*, 3, 2381–2391.

Braun, V., & Clarke, V. (2019). Reflecting on reflexive thematic analysis. *Qualitative Research in Sport, Exercise and Health*, 11(4), 589–597.

Carter, N. H., & Linnell, J. D. (2023). Building a resilient coexistence with wildlife in a more crowded world. *PNAS Nexus*, 2(3), pgad030.
Carus, P. (1898). On the philosophy of laughing. *The Monist*, 8(2), 250–272.
Curtin, S., & Kragh, G. (2014). Wildlife tourism: Reconnecting people with nature. *Human Dimensions of Wildlife*, 19(6), 545–554.
Davila-Ross, M., & Palagi, E. (2022). Laughter, play faces and mimicry in animals: Evolution and social functions. *Philosophical Transactions of the Royal Society B*, 377(1863), 20210177.
Despret, V. (2016). *What Would Animals Say if We Asked the Right Questions?* (B. Buchanan, Trans.). University of Minnesota Press.
Dunbar, R. I. (2022). Laughter and its role in the evolution of human social bonding. *Philosophical Transactions of the Royal Society B*, 377(1863), 20210176.
Ginzburg, J., Mazzocconi, C., & Tian, Y. (2020). Laughter as language. *Glossa: A Journal of General Linguistics* (2021–…), 5(1), 1–51. https://doi.org/10.5334/gjgl.1152
Gorichanaz, T. (2020). *Information Experience in Theory and Design*. Bingley, UK: Emerald Publishing.
Gorichanaz, T. (2026). Information experience for the expanding circle: Insights from animal–computer interaction and more-than-human–computer interaction. In N. Solhjoo (Eds.), *Multispecies Information Science* (pp. 40–52). Routledge.
Goumas, M., Boogert, N. J., Kelley, L. A., & Holding, T. (2022). Predator or provider? How wild animals respond to mixed messages from humans. *Royal Society Open Science*, 9(3), 211742.
Haraway, D. (2008). *When Species Meet*. University of Minnesota Press.
Kari, J., & Hartel, J. (2007). Information and higher things in life. *JASIST*, 58(8), 1131–1147.
Kline, C. S., Hoarau-Heemstra, H., & Cavaliere, C. T. (2023). Wildlife equity theory for multispecies tourism justice. *Journal of Travel Research*, 62(6), 1167–1180.
Mitchell, R., & Sinkhorn, K. (2014). Why do people laugh during dog–human play interactions? *Anthrozoös*, 27(2), 235–250.
Panksepp, J. (2005). Beyond a joke: From animal laughter to human joy? *Science*, 308(5718), 62–63.
Perry, G., Boal, C., Verble, R., & Wallace, M. (2020). “Good” and “bad” urban wildlife. In F. Angelici, & L. Rossi (Eds.), *Problematic Wildlife II: New Conservation and Management Challenges in the Human-Wildlife Interactions* (pp. 141–170). Cham: Springer. https://doi.org/10.1007/978-3-030-42335-3_5
Rychlowska, M., McKeown, G. J., Sneddon, I., & Curran, W. (2022). The role of contextual information in classifying spontaneous social laughter. *Journal of Nonverbal Behavior*, 46(4), 449–466.
Solhjoo, N. (2025). The multispecies perspective in library and information science. *Information Research*, 30(COLIS), 323–330. https://doi.org/10.47989/ir30CoLIS51898
Solhjoo, N. (2026). Research pathways for multispecies information science. In N. Solhjoo (Eds.), *Multispecies Information Science* (pp. 91–110). Routledge.
Taylor, N. (2011). Anthropomorphism and the animal subject. In R. Boddice (Ed.), *Anthropocentrism* (pp. 265–280). Leiden, The Netherlands: Brill. https://doi.org/10.1163/ej.9789004187948.i-348.61
Tinto, F., & Ruthven, I. (2016). Sharing “happy” information. *JASIST*, 67(10), 2329–2343.
Valeri, R. M. (2006). Tails of laughter: A pilot study examining the relationship between companion animal guardianship (pet ownership) and laughter. *Society & Animals*, 14(3), 275–293.

van der Linden, D. (2021). Interspecies information systems. *Requirements Engineering*, 26(4), 535–556.
Van der Linden, D., Davidson, B. I., & Gowland, S. (2025, June 23). Memorable urban wildlife encounters across the UK. Retrieved from https://osf.io/bd72q
Warren, C., & McGraw, A. P. (2016). Differentiating what is humorous from what is not. *Journal of Personality and Social Psychology*, 110(3), 407.
Winkler, S. L., & Bryant, G. A. (2021). Play vocalisations and human laughter: A comparative review. *Bioacoustics*, 30(5), 499–526.
Wolfe, C. (2010). *What Is Posthumanism?* Minneapolis, MN: University of Minnesota Press.

9

URBAN DOG PARKS AS INFORMATION GROUNDS

A Multispecies Family Perspective

Bhuva Narayan

Introduction

With the exception of a few months when relocating from one country to another, I have always had a dog or two in my home for the past 48 years of my life. First in India through my childhood and adolescence, then in the United States of America for 17 years, and now in Australia for the past 18 years. In essence, I have always lived in a 'multispecies family' (Irvine & Cilia, 2017) for the majority of my life. Being an introvert, a bibliophile, and a so-called creature of habit, my dogs, my fellow creatures of habit, have offered me an organic outreach into a third space outside of my home and my workplace. They have helped me form human–human friendships, while they themselves form their own unique, more-than-human relationships and friendships with other dogs and with other humans. Such a 'more-than-human perspective also opens up possibilities for understanding domains beyond family. [and] acknowledge that social life is always lived in more-than-human contexts, [for] nonhumans create its possibilities' (Irvine & Cilia, 2017, p. 8). In essence, similar to Gorichanaz's (2026, Chapter 3) in this volume, this work is a labour of love in the spirit of the liberal arts tradition and not an empirical or scientific research project in the traditional sense.

Unless one lives on a country farm or acreage, having a dog means walking your dog in your neighbourhood, in nearby nature trails, or along river or ocean beaches several times a day, and if one is lucky enough to have a designated dog park nearby, they may also have off-leash spaces, or dog exercise areas, for one's dogs to freely play and mingle with other dogs and humans in legally designated areas, fenced or unfenced. Thus,

DOI: 10.4324/9781003583424-11

> Dog parks not only provide insight into canine behavior but also into human-animal and human-human interaction. While such off-leash dog parks may appear simply as urban playgrounds for dogs, the interactions that take place there have implications that extend far beyond the fence that defines their boundary.
>
> *(Jackson, 2012. p. 270)*

Australia has a population of 26.5 million people and over 6.3 million domestic dogs (Australian Bureau of Statistics, 2022), almost half of all household pets. In my neighbourhood, 'dogs outnumber children 1.7 to 1, and with 1,295 dogs per square kilometre, the Inner West Council of the city, where Callan Park is located, holds the title for Sydney's highest canine concentration' (Iacono, 2024). Sydney has 370 off-leash places (Proctor, 2017), of which over 40 are in my Inner West Council area.

Without going into the politics and morality of multispecies domestic companions such as dogs, we shall proceed with DeGrazia's conditions to justify keeping animals in one's home: '(1) the animal's basic physical and psychological needs must be met, and (2) the animal must be provided with a life that is at least as good as she would likely have in the wild' (DeGrazia, 2002, p. 82). And that is where the role of dog parks comes in, especially in urban areas.

There is much literature on the multispecies aspects of urbanism in cities, with their socio-spatial dynamics expressed through often heated discourse around parks where dogs are walked, thus making dog parks some of the most contested spaces in any urban setting (Graham & Glover, 2014). However, there is not much literature on the everyday information interactions of humans and dogs in the context of dog parks, apart from Niloofar Solhjoo's works on multispecies families' everyday life (Solhjoo, 2024a, 2024b; Solhjoo et al., 2024) that touches briefly on dog parks.

In this chapter, I focus my observation on an urban park with several 'dog off-leash recreational areas' (Gomez, 2013) to conceptualise its affordances and map them to the information grounds theory originally conceptualised by Karen Fisher née Pettigrew (1999) when she studied a podiatrist's waiting room where the people waiting interacted with each other. Here I use the concept of affordance as originally described by Gibson (1979), as in 'the *affordances* of the environment are what it *offers* the animal, what it *provides* or *furnishes*, either for good or ill. … It implies the complementarity of the animal and the environment' (Gibson, 1979, p. 119), thus creating an ecological niche for both humans and their dogs.

The information grounds theory posits that some spaces 'foster information sharing in spontaneous, serendipitous, and planned ways' (Fisher & Naumer, 2006, p. 93), where people gather for one specific reason but start talking to others because other people are there. Similarly, walking into a dog park with one's dog instantly creates something that can be described as an 'information landscape' (Lloyd, 2006), which is socially constructed and imbued with meaning and is an inter-subjective space where one enters into a relationship with other humans and dogs sharing the space.

The original theory of information grounds (Pettigrew, 1999) has since been extended to various spatial contexts, both to physical spaces (Fisher, Landry, & Naumer, 2006; Savolainen, 2009), mobile digital networks (Counts & Fisher, 2010), online digital spaces (Narayan, 2013), and more recently, to a study of vehicle residents, people whose only home is their car and hence need to continually keep moving to various parks within the city due to local restrictions and regulations (Montague, 2024). In this chapter, I extend Karen Fisher's information grounds theory to postulate that off-leash dog parks are not only rich multispecies information grounds but also function as a node within a network of other such spaces within the city and beyond, through the interesting web of people and their dogs.

In this chapter, my main aim is to provide a rich description of the multispecies environment of a dog park within an urban setting and reflect on its attributes as it relates to how this park is transformed into information grounds for the people in its neighbourhood and nearby neighbourhoods. Although this is an *in-the-wild* ethnographic study, it is not a planned study with a research question, objectives, and findings, but a reflective retrospective analysis of my own notes based on daily observations, journal entries, and photos I have taken during the course of ten years (Figure 9.1).

FIGURE 9.1 My dog Gray: With one ear up, he's a good listener and knows how to keep a secret behind the other ear. (Photo by author, 14 November 2024).

The People-Place-Information Triad

Karen Fisher (née Pettigrew) originally coined the term 'information grounds,' which describes a people-place-information triad (Talip et al., 2016), or a synergistic 'environment temporarily created when people come together for a singular purpose but from whose behaviour emerges a social atmosphere that fosters the spontaneous and serendipitous sharing of information' (Pettigrew, 1999, p. 811). The concept of information grounds, as positioned within information behaviour research, was built on previous research in the field of sociology, specifically *the strength of weak ties* (Granovetter, 1973) and the concept of *third place* (Oldenburg, 1999). The strength of weak ties explains how people in our networks with whom we may have the weakest ties are the ones that act as connecting bridges to new networks that we may not otherwise have access to (Granovetter, 1973); this comes into play in dog parks, for our strongest ties in the dog park are actually our dogs, whereas the humans in the park are our weak ties, but over time, they expand our network. The concept of *third place* is of places outside of one's work or home where we come together to share an activity or simply hang out, and these spaces are open and inviting, comfortable and informal, convenient, unpretentious, have regulars, are often free, but often also have food and drink, and where light-hearted conversation is the main activity (Oldenburg, 1999). Although third places have been conceived basically as human spaces, dog parks are third places in the life of a 'humandog collective' (Hodgetts, 2017, p. 83); they can have the exact same elements as a third space for humans, including food and drink, which have increasingly become common with dog-friendly cafes and pop-up coffee shops in and near dog parks. For example, we have one named *Cafe Bones* that serves Puppacinos and Pup Muffins for the dogs, alongside coffee and breakfast for the humans. My dog Gray and his friends even know where to wait to 'pick-up' his Puppacino order once we pay for it.

Other Information Science studies that the information grounds theory builds on are *information in context* (Dervin, 1992); a series of information theories around the concept of *small worlds* (Chatman, 1999); and *life in the round* (Chatman, 1999), which posits that the 'insider' and 'outsider' structure and social norms of any given 'small world' lead to normative behaviour and to 'a public form of life in which things are implicitly understood' (Chatman, 1999, p. 212). According to Fisher (2006),

> As people gather in an information ground, they engage in social interaction, conversing about life, generalities, and specific situations that lead to serendipitous and sometimes purposive, formal and informal sharing of information on various topics…In terms of the temporal sense, as members disperse, so does the information ground, at least until the next gathering.

> Meanwhile, new situations arise in people's lives, local and global events continue to occur, and people pick up new information to be shared at their next information ground.
>
> *(Fisher, 2005)*

To the basic concept of information grounds, Fisher et al. (2004) added the following prepositions:

1. Information grounds are attended by different social types, most, if not all, of whom play expected and important, albeit different, roles in information flow.
2. People gather at 'information grounds' for a primary, instrumental purpose other than information sharing.
3. Social interaction is a primary activity at 'information grounds' such that information flow is a by-product.
4. People engage in formal and informal information sharing, and information flow occurs in many directions.
5. Information grounds can occur anywhere, in any type of temporal setting, and are predicated on the presence of individuals.
6. People use information obtained at 'information grounds' in alternative ways and benefit along physical, social, affective, and cognitive dimensions.
7. Many sub-contexts exist within an 'information ground' and are based on people's perspectives and physical factors; together, these sub-contexts form a grand context (Fisher et al., 2004).

As for definitions of what information is, many definitions of information end up being tautological, and there is no agreed-upon definition, as it depends on the context of its creation and use. In terms of defining information in context, there are some key uses of information:

a. 'Information-as-process'—the action of telling someone the fact of something (Buckland, 1991).
b. 'Information-as-knowledge'—knowledge communicated concerning some fact (Buckland, 1991).
c. 'Information-as-thing'—attributed to data and documents being informative (Buckland, 1991).
d. 'Information as a commodity'—related to the concept of its value as information progresses through the various steps of creation, processing, storage and use (Rowley, 1998).
e. 'Information as a constitutive force in society'—information as not just being embedded within a social structure, but also as an agent in the creation of that structure (Braman, 1989).

All of these contextual definitions of information come into play in the context of an urban park in Sydney, where the information practices of people with dogs result in an 'emphasis on social practices, the concrete and situated activities of [the] interacting people, [which are] reproduced in routine social contexts across time and space' (Savolainen, 2007).

In a dog park setting with people standing around watching their dogs play, there are also phatic expressions, or 'small talk,' that don't often directly convey any information, although they are an essential part of conveying context and serve a very important social function: 'through seemingly personal but inconsequential small conversation[s]; they build engagement, solidarity, and ceremony (habits) that create interpersonal trust through familiarity and interaction that is a crucial element of civil society' (Narayan, 2013).

In this chapter, I detail how a neighbourhood urban green park in Sydney, Australia, has become a living and thriving information ground for myself and my family, including my dog, and for everyone who frequents the park, and these dynamic information grounds often extend beyond the geographic location of the park itself, creating a network of information grounds across several dog-friendly spaces in the city, along a green corridor named The Greenway through expanding our collective 'information horizons' (Sonnenwald, 1999), and through enabling 'incidental information acquisition' (Williamson, 1998) or serendipitous or accidental discovery of information through encountering unexpected information (Erdelez, 2005).

A Place-Based Approach to Information Experience

As Marcia Bates asserts, 'there will continue to be several approaches in LIS to studying the phenomena of interest to our field' (2005, p. 8). To study the phenomena of information grounds within the context of dog parks, I use a place-based approach that includes a walking methodology and sensory ethnography. Often used by researchers in human geography (Massey, 2013; Tuan, 1977; Vanni & Crosby, 2023), it foregrounds the heterogeneity of place. I use my daily observations of the flora, fauna, humans, and dogs of my neighbourhood park, combined with my ethnographic interactions with them. I have maintained a diary of many of these interactions over the past ten years, combined with a daily photo record, which was my way of getting to know and understand my new neighbourhood when I moved here over ten years ago. This is complemented by my cumulative daily impressions based on sensory ethnography (Atkinson et al., 2008), which is an experiential process through which knowledge is produced, a 'process of learning through [...] multisensory, emplaced experience' (Pink, 2015, p. 64). This sensory approach includes the interconnected and interrelated aspects of smell, taste, touch, and vision that are all integral to the experience of any place.

Daily walks to my local dog park were the main data source for this study, which foregrounded the embodied nature of myself as a researcher, not to mention the bounding and dogful nature of my dog that directed some of my actions, and the directions and rhythms of my walks, which in my case, were mainly directed by my dog's own free will. According to O'Hare (2024), 'dog walks are a vital dimension of everyday place-making, with the dog, the human, and their symbiosis instrumental to the mediation of public space. Indeed, it is this 'everydayness' of dog-walking that makes it ripe for further exploration' (p. 1313).

Bates (2002) emphasises both active and passive modes of information seeking, for 'it is not unreasonable to guess that we absorb perhaps 80 percent of all our knowledge through simply being aware, being conscious and sentient in our social context and physical environment' (p. 4). Much of the information I gathered from the dog park was not actively sought for research purposes but was acquired through passive imbibing of the information in my environment; hence, this study is *a posteriori* as such. As someone who literally stops and smells every flower, touches plants to feel them, occasionally tastes unidentified wild berries, and takes photos to remember things by, I have amassed a huge archive of sensory provocations from my walks, not unlike my dog's own sensory explorations. Through this approach, I was documenting my personal experience for my own memory, rather than seeking information, either actively or passively.

As senses are a way of absorbing information, one might say that I was engaged in an information experience, for 'information experience is a person's in-the-moment engagement with information, and it depends on what a person pays attention to and finds informative' (Gorichanaz, 2020, p. 25); however,

> The word 'experience' broadly has two senses: on one hand, it can refer to a slice carved out from a person's first-person perspective of the flow of existence, while on the other hand the word can refer to an accumulation of the flow of existence.
>
> *(Gorichanaz, 2020, p. 24)*

In many ways, my daily experiences over ten years fall under both categories of experience, the specific and the cumulative. It was possible only through what Anna Tsing (2015) calls the 'art of noticing,' 'but it's noticing on a mission. It's noticing that allows a researcher to say something general about an object of study' (Rutherford, 2020).

Over time, I have observed other rhythms that align with or intersect with my rhythms. According to Springgay and Truman (2018), the four major themes in walking research are place, sensory inquiry, embodiment, and rhythm, where 'walking becomes a way of inhabiting a place through the lived experience of movement. Walking is a way of becoming responsive to place; it activates modes of participation that are situated and relational' (Springgay & Truman, 2018, p. 4). Place performs an integral role in walking research as it is imagined as a particular location, activity, or incident… "by walking in a place one can activate modes of participation that are situated and relational as walking can connect bodies, environment, and the sensory surrounds of a place" (Springgay & Truman, 2018, p. 27). This connects well to the place-based approach, for the 'human tendency [is] to locate information spatially' (Miller, 1993, p. 3). And within any space, 'we can only experience that which we attend to…what a person experiences as informative' (Gorichanaz, 2020, p. 24).

Although I background this study through the scholarly literature in urban sociology and human geography, I analyse my observational data through an Information Science perspective, mapping ecological concepts to Information Science, specifically to information grounds (Fisher et al., 2006) within a more-than-human multispecies information sphere. However, the limitations here are that whatever I think or write is going to be perceptually anthropocentric, for I have no direct knowledge of my dog's *umwelt* or his perceptual life-world (von Uexküll, 2013), except through what I perceive are his responses to the environment. As Bates (2026, Chapter 2) reminds us, we need to be aware of [our] significant differences in (perceptions and senses) when studying interactions between humans and other animals. My own *umwelt* consists of observing and responding to the physical ecology around me at all times, for my own 'sensemaking' (Dervin, 1992), and so does my dog's, although we respond to different things within it, based on our different sensory modes and different *unwelts*. As discussed in Chapter 2 by Marcia Bates, both humans and nonhumans are loci of intense information processing shaped by their unique nervous systems, communication modes, and adapted behaviour settings—environments we inhabit that are structured to support the activities we engage in and share. Most of my interactions with my dog are therefore information below the surface (Gorichanaz, 2026). His olfactory clues, while not experienced directly in all their complexity by me as a human, are instead made available to me at an indexical level through communicative signs that retain material links to their cause (Hodgetts, 2017), such as Gray tracking the path of another dog through scent alone while leaving his own scent for others to note. This human–animal connection is reciprocal and is a form of interspecies sensemaking; humans learn from animals, and animals learn from humans (Grenersen, 2016), and my own human–dog information interactions are characterised by love, living, and learning and generate a unique web of interactions and shared experiences, encompassing embodiment, communication, cognition, empathy, and social engagement (Solhjoo, 2024c).

Privacy and Other Ethical Considerations

Privacy is one of the 15 propositions within the people-place-information trichotomy in the literature on Information Grounds, for

> The level of perceived privacy affects people's preference for particular information grounds. Places that include private areas for talking or tables that are far enough away from each other foster conversations that may be personal. Conversely, information grounds might be attractive because they enable eavesdropping, which may contribute to the overall richness of the place.
>
> *(Fisher et al., 2006)*

The concept of privacy is also relevant in terms of my methodology. First, ethical clearance (UTS Ethics #ETH24-981) was obtained from my university's ethics review board to write up this research based on collecting data through unobtrusive observations in a public place, as no interviews or surveys were conducted. No individual dog or dog owner or park visitor is named or identified in any way, and pseudonyms are used where individual dogs and people are mentioned. Although I see the people and their dogs in the park every day, I do not know any of them personally and only know them through their dogs' names.

Susan Orlean writes about 'the unwritten rules of the dog park' (Orlean, 2011) in New York City, and they hold true in my dog park too, especially her first and last points:

1. Ask the dog's name, never the owner's name.
2. For some reason, every single request for the name of the dog is immediately followed by an inquiry about the age of the dog. I have no idea why, but, believe me, this is true.
3. Compliment every dog on its beauty. If the dog is one-eyed, three-legged, gap-toothed, and mangy, make your compliment more fulsome. Having trouble thinking of a way to effuse about a pug-ugly mutt? Just remember, no one ever tires of hearing how "great" a face their dog has. Remember: "great" sounds really good without having a specific definition.
4. Being deeply engaged in a cell-phone conversation while your dog is humping another dog, gnawing off someone's leg, or terrorizing a puppy is generally considered "anti-social" and "not cool."
5. Dog parks are more cliquish than any other human gathering with the possible exception of seventh grade. Deal with it.
6. In an interesting inversion of status, the reigning breed in the dog park these days is the really-oddball-unidentifiable-mixed-breed-mutt-found wandering-the-street or its equivalent. The stranger the mutt, the better;

the more peculiar the circumstance of it coming into your life, the better. If you have a purebred dog, it better be a rescue, or you're going to be sneered at as an unbearable snob with questionable morals. (I say this as an observation about an encouraging social trend, not a criticism, by any means).

7 You may never learn the names of any of the people you talk to in a dog park, even after many, many hours spent there with them, and many hours of conversation. But if—knock on wood—anything should ever happen to your dog, these nameless non-strangers will rally, sympathize, offer to help, and hold your hand.

(Orlean, 2011)

This is in line with codes of public interaction in the modern world, where 'everyday civility requires that strangers maintain a sense of separateness in crowded places' (Wise, 2008, p. 10). In the context of dog parks, this applies to both physical and mental interactions between humans, not to mention dog–dog and dog–human interactions. However, it is indeed true that these no-name non-strangers in my park will readily come to my aid if I ever need anything dog-related.

It is analogous also to Elfreda Chatman's concept of *life in the round*, for dog parks create a form of 'life in the round, a public form of life in which things are implicitly understood' (Chatman, 1999, p. 212), where members are concerned with the creation and support of social roles within that world where a certain routine is followed and their social roles of insider and outsider set boundaries of behaviour, and where people do not cross information boundaries unless deemed as critical (Chatman, 1999). This can also be said of dogs in a dog park and the dog–animal context studied here, for dogs, like us, are social animals, from the wiring of our brains to the shape of our societies, always looking to read the minds of our fellow animals. Although animals may not consent to being part of any research in the way humans can (Solhjoo, 2024c), they can express their choices and wishes through their behavioural cues, which I can only interpret through my own limited human cognitive frame and help them enact their choice; for example, I will choose to go in a certain direction on my walk if Gray indicates that he wants to have a swim, or in another direction if he wants to go and play with his friend Princess, the French Bulldog, instead. For readers who have dogs, you will already know how dogs express their intentions, joy, fear, and desires through eye contact, pointing, nosing, and through several other subtle cues.

A Multispecies History of the Place

A seemingly empty space like Callan Park, an urban green park—'relative location within a bounded space' (Tuan, 1977, p. 13)—slowly transforms itself into a place—'a center of meaning constructed by experience' (Tuan, 1977,

p. 151)—either physically or conceptually through our own constructions of a sense of place, rather than just the structure of space (Harrison & Dourish, 1996). In essence, 'humans layer their own understandings onto abstract space in order to create subjective places' (Jones & Evans, 2012. p. 2319).

This sense of place does not have to be discrete, as in *a* place, but may be generated anywhere through movement between, along, and through different places (Ingold, 2011). To feel a sense of place, the actual place (as in a space) itself is not crucial, for inhabiting a place is a process. A place is not just a home we inhabit, but the context of the home—the roads that lead to and out of one's home, and familiar things and the information we gather along the way to the dog park or the shops. As Ingold says, 'habitation is lineal. That is to say, it takes people not *across* the land surface but *along* the paths that lead from place to place' (2011, p. 34). Thus, a place is a palimpsest of meanings (Bailey, 2007, p. 208) and a living archive (Cresswell, 2012) of lives that have passed through there. And hence it is important to understand the public history of any place we inhabit.

Habitation

The City of Sydney is 'made by intersecting trajectories, mapping onto place the British Empire, colonialism, Indigenous knowledge and histories, industrial development, post-war migration to Australia, environmental change, and urban renewal' (Vanni & Crosby, 2023, p. 4). Traditionally, the lands of the Cadi and Wan people of the Eora nation, commonly referred to as Cadi-gal and Wan-gal lands, although an Indigenous elder recently shared that the suffix -gal refers to the men of the land and hence was a colonial patriarchy-influenced tradition (Foster & Manning, 2023).

On my frequent wanderings in Callan Park with my dog, I walk past several of the 'public art as wayfinding' installations from the Council's Gadigal-Wangal Wayfinding Project showcasing Indigenous artists, and I am continually inspired by their connection to the place and its stories, stories that reveal 'how we human beings are related, not only to each other, but to all things, animals, plants, rocks—in fact, to the very stuff the stars are made of. This relationship is a kinship relationship' (Harris & Wasilewski, 2004, p. 492). One can see some of these Indigenous artworks that embody multispecies stories and perspectives while walking in the park. My favourite one is titled *Fish Traps* by Edward Clarke (Figure 9.2) and stands as a remembrance of local food practices in the past.

Landscape

The *Callan Park Conservation Area & Buildings* has a more recent colonial history (1884–1994) as the site of a mental health facility predicated on garden

FIGURE 9.2 Fish traps by Heather Clarke and Chris Edwards. (Photo by author, March 26, 2022).

therapy, then a pretty rigid psychiatric hospital for 'the insane,' a now-infamous lunatic asylum. Today, Callan Park, a 60-hectare urban waterfront park, is a recreational park and conservation site with Indigenous habitation evidenced through five 4,500-year-old shell middens. The park is filled with over 130 human-made structures, many of them historic buildings, some disused, many dangerously falling apart and cordoned off, which are now filled with feral cats lounging in the eaves. There are people who feed the cats, and there are others who work with the *Cat Protection Society* to spay/neuter and home them. The buildings are surrounded by remnants of a classical garden, a therapy garden, bushlands, wetlands, salt marshes, sporting fields, dog off-leash areas, dog beaches, various pontoons and jetties, a boathouse, serenity gardens, a community farm, outdoor gyms, and a children's park. Its waterfront boardwalks are part of an 8-mile running circuit that extends beyond the park, named the Bay Run. It is a heritage-listed site within the Inner West Council, an inner-city council on the western edge of Sydney's City Business District (CBD or downtown).

Callan Park is surrounded by residential neighbourhoods on all sides except the waterfront, and there is generally ample free parking all around, so people come from other areas of Sydney to walk their dogs here too, especially on weekends. The park has been used for off-leash dog walks for decades, and during the long COVID-19 pandemic lockdowns in Sydney in 2020–2021, more and more people started using the park, and when the Greater Sydney Parklands Authority was created in 2022, the ownership of Callan Park was transferred to a trust (Centennial Park and Moore Park Trust—CPMPT), which warned dog owners that off-leash dogs are illegal and may cop a fine. However, the local council announced they would not enforce the ban as there have been no major incidents with off-leash dogs; the park rangers are quite friendly and helpful in case anyone needs help. Meanwhile, park use by dogs and their companions has only increased as more and more people work from home since 2020.

People

During the day, Callan Park is abuzz with runners, mothers and fathers with young children in prams, toddlers (some on leashes!), alongside cyclists, athletes, school groups, sports groups, and dog walkers. There are groups of people practising yoga, laugh therapy, writers hurrying to events at the NSW Writers' Centre, photographers, kids at a bush school walkabout, and art school students with their drawing books or easels. Weekends mean the park is packed with sportspeople in the different grounds, playing cricket or football, and with kids practising athletics. This also means the parents are waiting around on a picnic blanket on the periphery, often with their dogs, and some of them occasionally get the ball.

There are separate groups of people that recognise each other every day—those who walk their dogs once a day after work, those who walk theirs twice a day, early in the morning and after work, and those who walk their dogs three times a day—these are often those who work from home, and hence their dogs nudge them out of home whenever they can. These are all the people that this chapter deals with, especially around their everyday information experience in the dog park and how it constitutes an information ground.

Flora and Fauna

The park has many major native species through habitat restoration over the years (Greater Sydney Parklands, 2025). The park is also home to the native golden wattle *Acacia mearnsii*, or Black Wattle to the Indigenous people (Figure 9.3). The state-level Callan Park [Special Provisions] Act (NSW Government, 2002) was enacted to maintain the park in public ownership, with any development subject to public control, which has staved off commercial developers so far.

Every morning a noisy conference of the birds convenes—even from my home two streets away, I can hear Kookaburras, Miner Birds, Koels, and

FIGURE 9.3 *Acacia mearnsii*, or golden wattle, with a sulphur-crested cockatoo watching over Callan Park. (Photo by author, September 11, 2022).

Currawong, and evenings are when the parliament of birds holds court—Cockatoos, Corellas, Rainbow Lorikeets, and Australian Magpies, with the Australian Ibises and the Crested Pigeons raiding the day's detritus on the grass. Closer to the water, the Seagull orchestra plays, and the occasional Cormorant and Pelican can be seen on rocks jutting out of the water, drying their enormous wings.

During the day, the park is home to seemingly happy dogs of all kinds: Yappy, snappy, waggy, shaggy, and often soggy dogs emerging from the beach or a swim in the bay (Figure 9.4). Other dogs strut along the dam in the bay (Figure 9.5). Sometimes there's a small dog in a pram, or an elderly dog in a beach cart, or a disabled dog in prosthetic wheels, and almost every day, a very frisky three-legged dog who is completely unaware he is missing a leg at all.

During the night, the park is a haven for owls, possums, and thousands and thousands of grey-headed flying foxes (also known as fruit bats, the largest members of the bat family by size) that visit the huge 200-year-old native Moreton Bay Fig trees that number in the hundreds and line the park. On a sunny day, lizards and water dragons can be seen lazily sunning themselves near the ponds. An occasional snake sighting may cause quite a hullabaloo among the dog walkers.

FIGURE 9.4 Dogs on the beach. (Photo by author, 16 February 2024).

FIGURE 9.5 When the tide is low, one can walk on the dam by the bay. (Photo by author, 25 August 2018).

There is evidence that in pre-colonial times, everywhere in what is now Sydney was dingo territory (Hull, 2018). A dingo is 'an ancient breed of domestic dog that was introduced to Australia, probably by Asian seafarers, about 4,000 years ago. Its origins have been traced back to early breeds of domestic dogs in south east Asia' (Australian Museum, 2024). Aboriginal people in what is now Sydney, and all over, lived with and tamed dingoes and used them for hunting and for warmth. Dingoes were also part of the inner circle with humans and part of Indigenous Dreamtime stories (Australian Museum, 2023) and are represented as guardians and noble spirits. Today, Callan Park is bereft of dingoes. The introduced Red Fox is more common in the bush country north of the river from Callan Park, but is occasionally spotted in the park and reported to FoxScan (2025), a fox tracking site. One early morning, we had quite a bit of excitement in the neighbourhood when a neighbour walking a young golden retriever puppy reported that a fox came out of the woods and played with their puppy. And disappeared into the woods again.

I must note here that my chapter will not describe the behaviours or interactions with or between any other animals (apart from dogs) inside this park, for (1) the native species here are protected by law, and (2) people are advised not to directly interact with them.

The Contemporary Canine Environment

The Dogs

Although Elizabeth Marshall Thomas' highly celebrated book *The Hidden Life of Dogs*, released in 1993, has been criticised for its non-interventionist, laissez-faire approach to letting her dogs free roam the city in order to observe them in the urban jungle, which would not get past contemporary ethical review boards, one of her conclusions in answering the question: *What do dogs want most?* Is an emphatic *'Not food, not sex, but other dogs'* (Thomas, 2010, p. 280). This is something many of us dog lovers will completely agree with and go out of our way to help our dogs socialise with other dogs as often as they need to (Figure 9.6). For urban dogs, this means 'social integration into the larger group of pets and pet owners. From that perspective, dogs must be able to interact with other dogs and human beings and manage dynamic but socially controlled interactions' (Tissot, 2011, p. 275).

There is much literature on the multispecies urbanism of cities with their socio-spatial dynamics expressed through often heated discourse around parks where dogs are walked, thus making dog parks some of the most contested spaces in any urban setting (Graham & Glover, 2014). The historical turning point for some of the debates was the so-called Burnley Dog Wars in the UK in 1977, when 300 dog walkers, accompanied by their canine companions,

FIGURE 9.6 Gray, with his dog-walking friends, goes out every afternoon with a professional dog walker. (Photo by author, 20 July 2023).

assembled in opposition to a 'No Dogs Allowed' ruling enacted for Scott Park in Burnley. In writing about the Burnley Dog Wars, Pemberton (2017) states that

> Interspecies sharing of public space is not a straightforward expression of human care and obligations to dogs. It demonstrates the significant roles of culture, human priorities and ideas about risk in framing the rules by which dogs enter public space, rules that create and police dog-walking space.
>
> *(2017, p. 267)*

Dogs are policed in urban areas in the same way as graffiti is policed, as

There is also an argument that dog parks are a sign of gentrification and create social boundaries to exclude others, as well as define insiders, and that, through animals, middle-class residents have developed exclusive public spaces in gentrifying areas (Tissot, 2011). On the other hand, there is also evidence that dog parks have several benefits for dogs, humans, and the community (Gómez et al., 2018), including strengthening the social fabric and feeling of safety in a community (Wood & Christian, 2011), for 'dog parks are a resource for building social capital through interpersonal exchange…and the dog serves as a vehicle to accept our positive and negative emotions, and in return provide enduring camaraderie and support' (Vincent, 2019. p. 518).

The Porous Boundaries of the Dog Park

The main site of my observations over the past ten years is the above-mentioned Callan Park. It is a block from my home, and our dog Gray thinks that it is imperative that we go there three times a day on weekdays: In the morning with my partner, then midday with a professional dog walker and a few other dogs, and then again in the evening with anyone who'll take him there, and we do have a lot of takers; each one of these three walks lasts an hour or more. There is also a final 10-minute walk before going to bed. On weekends and holidays, there is no professional dog walker, so we take Gray for a drive and explore one of Sydney's 370 off-leash places (Proctor, 2017) to play—I call them our 'car walks.'

In 2025, after a lot of consultations with the dog park users and residents in the area, there are now designated off-leash and on-leash areas in the park (Figure 9.7), with the majority of the sporting fields (leased to the council) on 'time share,' i.e., dogs are welcome to play off-leash only when there is no registered sporting event. Absolutely no dogs can be off-leash in the park after dark, a rule to protect the native animals that rule the park at night.

At the Northeast, entrance to the park, and outside of the official Callan Park boundary (Greater Sydney Parklands, 2025), is a park named King George Park, a grassy oval sporting field with restrooms and sheltered barbeque and picnic areas, all owned and managed by the council, which has become one of the main 'timeshare' places for residents to congregate with their dogs, as it is closest to the residential neighbourhood and also to a free car park; however,

FIGURE 9.7 My dog Sammy checking out the dog off-leash etiquette. (Photo by author, 20 January 2020).

FIGURE 9.8 My dog Sammy walking along Balmain Shores with a view of Iron Cove Bridge. (Photo by author, 8 December 2019).

most people do not seem to differentiate between this park and the main Callan Park, as there are no visible barriers between them. There is also another private off-leash park to the east, just a minute's walk east along the boardwalk under the Iron Cove Bridge, that belongs to residents of a posh waterfront residential complex named Balmain Shores that welcomes non-resident dogs for play, so long as the dogs do not go inside the children's play area (Figure 9.8).

The NSW Government has held various public consultations and 'have your say' campaigns about Callan Park-related matters, including a proposal to build a 'tidal bath' off the waterfront and install artificial turf in some of the sporting fields, with some strong reactions from the dog lovers, who have raised many petitions together, all organised while having informal chats while their dogs play. Politicians often visit the park on the weekends and talk to the locals and visitors about the park and about environmental and other issues nationwide.

Such continual discourse around urban dog parks illustrates that

> the city is not solely a human habitat but is instead – whether visible or accepted – a place firmly inhabited by a variety of species, even if there has been little attention given to the ways in which specific human–animal configurations become politically contested.
>
> *(Urbanik & Morgan, 2013, p. 292)*

Notice boards everywhere along the waters inform people that the fish from the waters are not safe to eat due to the heavy metals in the water from when there was an industrial area by the waters, but anglers fish anyway, especially on weekends. Dog walkers are wary of the anglers, as they sometimes leave sharp, nail-like shrimp bits lying around.

The Information Environment

The Dogs and Their Walkers

Dogs are 'an antidote for the human anonymity of the public places of our contemporary society. Dogs facilitate contact, confidence, conversation, and confederation among previously unacquainted persons who might otherwise remain that way' (Robins et al., 1991, p. 23). However, this 'requires careful, conscious and very deliberate mediation and is subject to constant negotiation and renegotiation across spatial, social and temporal dimensions of their existence' (O'Hare, 2024, p. 1326).

As 'dog walking is a … co-produced practice, and…is performed within animated public spaces including streets, and footpaths, public highways and in parks and green spaces' (O'Hare, 2024, p. 1315), it also results in the need to socialise our dogs, which paradoxically is 'inseparable from a desire for dominance through which humans seek to subdue the unruly forces of nature' (Tuan, 1984, p. 5).

In the context of dog walking, we also seek to regulate human nature itself, for 'walks are exposed to an abundance of performative interactions with multiple, intersecting humans, wildlife, and other dogs and their walkers. They are also routinized and ritualized, with walkers often following familiar practices and routes' (O'Hare, 2024, p. 1315), with consequent rule-making about appropriate dog park behaviours. Within these human-dog-human expectations, dogs negotiate whatever freedom they have within the confines that humans assign them (Masson, 1998), despite the physical restraints of the mobile space they occupy (when on a leash) and the spatial restrictions imposed on them when off-leash. In the liminal space of the dog off-leash areas in parks, imparting relevant information about one's dog to others is key to avoiding any misunderstandings. For example, one dog always has a bandanna that says, 'Do not feed me!'

During the daytime, there are a lot of professional dog walkers who collect their clients' dogs in their cars (no more than four at a time) and bring them to the park for a play in the off-leash areas. A professional dog walker could easily end up walking a dozen dogs a day over 3–4 trips. There are over two dozen professional dog walkers who use the park, and they all help each other out and cover for each other as needed, for they are all in quite high demand. The dog walkers were initially met with hostility by so-called owner walkers, but over the past ten years, the dog walkers have become the most trusted park users, as they know every nook and corner of the park and disseminate information about any issues they see, and often also organise and advocate for the rights of all park users. They have evolved now into a powerful voice in consultations with the council about park-related decisions. Clients of the professional dog walkers such as myself also trust them completely, as they are better informed of the day-to-day issues in the park: They warn us of the council spraying fertiliser on the lawns, report poison traps left by the council for foxes or rat poison in food left by dog-haters (sadly, a reality), report any canine disease outbreaks, and take quick action on lost and found dogs (and even found objects) through the various information networks.

The unavoidable 'everydayness' of dog walking also means establishing a routine around one's work and other commitments, and yet it is also the part of everyday life that many dog lovers enjoy, albeit sometimes grudgingly, whether they are functional walks or recreational walks. This also means that whatever one's routine is, there are likely to be others with the same routine, and slowly but surely, one begins seeing the same dogs and humans every day, and we become part of a habitus, gaining some social capital with each other in the process (Gómez et al., 2018). This also leads to some cliques within dog lovers who inhabit the space at certain times; if one goes there outside of one's own routine, one may feel like an outsider. There are frisbee throwers, ball chuckers, football kickers, tug-of-war enablers, stick throwers, and dog-agility enthusiasts, and they do not always intersect or hang out with each other. Like Susan Orlean says in the *New Yorker*, 'Dog parks are more cliquish than any other human gathering with the possible exception of seventh grade. Deal with it' (2011). However, if a person is walking a friend's dog or dog-sitting for a local, the dog people will recognise the dog and welcome the guardian into their circle.

Phatic communication in dog parks often leads to more substantial and meaningful conversations over time, and often, one gets to know a lot of information about a person and their dog without ever knowing their name or occupation. Occasionally, when two dogs really get along and play well, their humans may even plan to meet on certain days and times, and after a few such friendly sessions, they may even exchange phone numbers. Even then, the phone number is just entered as Gray's dad, or Monty's mum, or similar. And the next thing you know, we're all attending dog birthday parties in the park. Sadly, also some dog memorial services (Figure 9.9).

Callan Park also has several community social media pages for information sharing: One run by the Inner West Council for dog-related matters, one by the F*riends of the Callan Park*, a volunteer community organisation that advocates to maintain the park as a heritage site, and one named *Callan Park Dog Lovers* (with over 2100 followers) run by some park users, some of whom also happen to be published writers who use the park for walks with their dogs and for

FIGURE 9.9 A small section of a large dog memorial space under a railway bridge in a dog park nearby. (Photo by author, 2 August 2024).

inspiration—one of them writes crime fiction, including a light-hearted series with dog detectives. And of course, there is a lot of information sharing about veterinarians, dog grooming, dog boarding, and doggie day care. Several of the dogs in the neighbourhood also have their own social media pages maintained by their humans that bring out the uniquely different personas (or dogsonas) of the dogs through photos, anecdotes, and stories; Gray has one, maintained by my partner, which has a unique perspective of a dog striving to understand his world in order to obtain a dogtorate by talking to various doctoral advisers—he has many friends and followers that interact with him online. Such interactions 'form a parasocial relationship with the animal, wherein … the audience does not physically interact with the animal but only with its mediated representation through companion species content' (Shamayleh & Arsel, 2025). Such online information interactions also get folded into the information grounds of the dog park, where we get to know our neighbourhood dogs (and a little bit about their humans) without having to be friends with their humans on social media.

Over the past several years, there have been some very successful dog meetups, starting with Golden Retrievers on the last Sunday of every month, with over 100 people gathering on the lawn with their Golden Retrievers and other dogs, some of them guide dogs in training. This trend has now resulted in other meetups on other weekends, including a Dalmatian meetup, a Ridgeback meetup, and, more recently, a pug meetup. All of these spontaneously create breed-specific information exchange. Nevertheless, the majority of the puppies and dogs that come to the park are crossbreeds that have been rescued from dog shelters, like my dog Gray. The social media pages for the dog shelters and adoption agencies are a major source of information for those looking to adopt or for those looking to foster. In fact, that is how we found Gray as a puppy, rescued with his siblings from a remote rural shelter. Many dog trainers use the park too and become instant sources of information, advice and wisdom for those who talk to them.

One remarkable aspect of this 'information ecology' (Nardi & O'Day, 1999), a metaphor to view any local information space as a complex information ecosystem that is a hybrid information space of online and physical space, is how everyone rallies together to help someone in need, especially for unforeseen dog-related expenses. If a dog has been in an accident or needs major surgery for some issue, a park friend makes a fundraising page with all the requisite information and posts it on all the park-related social networks, and the money is raised within days, even if it is to the tune of over 10000 dollars; I have seen this happen three times in the past five years. When one of the dog walkers met with an accident that totalled her dog van, the community instantly rallied to raise money for a new van.

The park generates other information grounds around parents, kids, sports, and even personal trainers and yoga teachers who use the park. There are several boats moored to the bay and also several dinghies stored in a dinghy bay next to the pontoon, and there is an information ground around them. There are other organised events in the park, such as walks and runs for certain causes, and also music festivals and other events, which bring intersections with information grounds from across the city. Halloween is a time for the park visitors to go ghoul hunting around the grounds of the old mental hospital, or on organised ghost tours, and also for dogs to come dressed up in costumes.

There have been a few commercial film shoots on the grounds of the park, which always generate some interest but also word-of-mouth information about what sections of the park to avoid while walking your dog (Figure 9.10).

On Saturdays, the timeshare ovals in the park are often booked for kids' sports, so most of us travel to other smaller parks around Sydney with our dogs, and this has resulted in our information grounds expanding into other neighbourhoods and the community getting involved with activism around issues such as 'Stop Greyhound Racing,' 'Stop Horse Racing,' and other issues around the need for humane treatment of animals. In 2017, hundreds and hundreds of dog lovers from across Sydney took our dogs to the greyhound racing track to protest greyhound racing. Although such racing is not officially banned as yet, there is a lot more scrutiny around the welfare of the greyhounds and the general conditions in this industry.

FIGURE 9.10 A golden retriever meetup in Callan Park. (Photo by author, 19 Dec 2021).

Conclusion

'As an emerging field, multi-species ethnography has opened up a new perspective on human-animal relationship and can better analyze the participation of animals in human society' (Xuanyu & Xiaoyu, 2023). There have already been a few studies of multispecies households in the past few years by Solhjoo and co-researchers (Solhjoo, 2024a, 2024b; Solhjoo et al., 2024), focusing on multispecies Information Science. This chapter extends it to highlight just the dog-human dynamics within the public commons of urban green spaces, especially dog parks. Dog parks, apart from becoming information grounds themselves, also connect with other information grounds to allow information flow and communication between several parks and other allied dog-related ecosystems. This has enabled a network of several different information grounds along a green corridor named The Greenway (Figure 9.11), although each one of them was temporarily created when people came together for the singular purpose of exercising their dogs, but from this practice emerged a social atmosphere that fostered the spontaneous and serendipitous sharing of information that contributes to the social fabric of the city. Thus, the public commons of the dog park have evolved to become information grounds for the community, and these grounds extend well beyond the confines of the space and place of the park itself.

The implications for this conceptualisation of dog parks as information grounds point to the importance of urban dog parks as valuable community spaces that bring neighbours and neighbourhoods together in positive ways and have implications for the design of urban neighbourhoods in both

FIGURE 9.11 Map of the greenway. (Photo by author, 12 February 2025).

high-density housing and suburban areas. They need to be considered as essential urban infrastructure, just like green spaces for everyone, play areas and children's playgrounds.

Acknowledgements

I would like to acknowledge that I wrote this chapter from my current home located in the ancestral lands of the Wan people of the Eora Nation by the Parramatta River and pay my respect to their Elders, both past and present, acknowledging them as the traditional custodians of knowledge for this land, its waters, and its skies, and whose stories and wisdom inspire me every day.

The chapter is also inspired by the oral storytelling of my partner, Dr Friendless, a fellow dog whisperer, and by my two most recent companion dogs, Sammy (2003–2020), a Kelpie-cross rescue, and Gray (2023–present), an Australian Cattle Dog rescue, without whose existence and daily insistence these observations would not have been possible. Both Sammy and Gray have their distinct doggie personas, or 'dogsonas,' that have elicited uniquely different ways in which their humans interact with their worlds.

References

Atkinson, P., Delamont, S., & Housley, W. (2008). *Contours of Culture: Complex Ethnography and the Ethnography of Complexity*. Rowman: Altamira.

Australian Bureau of Statistics. (2022) Total number of pets owned in Australia from 2019 to 2022, by type. Report. *Statista Research Department*. Accessed from https://www.statista.com/statistics/1294587/australia-total-number-pets-owned-by-type/

Australian Museum. (2023). First Nations storytelling. *Australian Museum*. Accessed from https://australian.museum/publications/birds-storybox/first-nations-storytelling/

Australian Museum. (2024). Dingo. *The Australian Museum*. Accessed from https://australian.museum/learn/animals/mammals/dingo/

Bailey, G. (2007). Time perspectives, palimpsests and the archaeology of time. *Journal of Anthropological Archaeology*, *26*, 198–223. https://doi.org/10.1016/j.jaa.2006.08.002

Bates, M. J. (2002). Toward an integrated model of information seeking and searching. *New Review of Information Behaviour Research*, *3*(1), 1–15.

Bates, M. J. (2005). An introduction to metatheories, theories, and models. In K. E. Fisher, S. Erdelez, & L. McKechnie (Eds.), *Theories of Information Behavior* (pp. 1–24). Medford, NJ: American Society for Information Science and Technology, ASIS&T Monograph Series.

Bates, M. J. (2026). Researching information behaviour among multiple species. In N. Solhjoo (Eds.), *Multispecies Information Science* (pp. 22–38). Routledge.

Braman, S. (1989) Defining information: an approach for policymakers, *Telecommunications Policy 13*(3), 233–242.

Buckland, M. K. (1991) Information as thing. *Journal of the American Society for Information Science*, *42*(5), 351–360.

Chatman, E. A. (1999). A theory of life in the round. *Journal of the American Society for Information Science*, *50*(3), 207–217.

Counts, S., & Fisher, K. E. (2010). Mobile social networking as information ground: A case study. *Library & Information Science Research*, *32*(2), 98–115.
Cresswell, T. (2012), Value, gleaning and the archive at Maxwell Street, Chicago. *Transactions of the Institute of British Geographers*, *37*, 164–176. https://doi.org/10.1111/j.1475-5661.2011.00453.x
DeGrazia, D. (2002). *Animal Rights: A Very Short Introduction*. Oxford: Oxford University Press.
Dervin, B. (1992). From the mind's eye of the user: The sense-making qualitative-quantitative methodology. In J. Glazier & R. R. Powell (Eds.), *Qualitative Research in Information Management* (pp. 61–84). Englewood, CA: Libraries Unlimited
Erdelez, S. (2005). Information encountering. In K. E. Fisher, S. Erdelez, & L. McKechnie (Eds.), *Theories of Information Behaviour* (pp. 179–184). Medford, NJ: Information Today, for ASIS&T.
Fisher, K. E. (2005). Information grounds. In K. E. Fisher, S. Erdelez, & L. McKechnie (Eds.), *Theories of Information Behavior* (pp. 185–190). Medford, NJ: Information Today, Inc.
Fisher, K. E., Durrance, J. C., & Hinton, M. B. (2004). Information grounds and the use of need-based services by immigrants in Queens, NY: A context-based, outcome evaluation approach. *Journal of the American Society for Information Science & Technology*, *55*(8), 754–766.
Fisher, K. E., Landry, C. F., & Naumer, C. (2006). Social spaces, casual interactions, meaningful exchanges: 'information ground' characteristics based on the college student experience. *Information Research*, *12*(2) paper 291. Accessed from https://informationr.net/ir/12-2/paper291.html
Fisher, K. E., & Naumer, C. M. (2006). Information grounds: Theoretical basis and empirical findings on information flow in social settings. In A. Spink & C. Cole (Eds.), *New Directions in Human Information Behavior* (pp. 93–111). Dordrecht: Springer Netherlands.
Foster, S., & Manning, N. (2023) 16 November. *Working with Community to Create Indigenous Perspectives on Object Based Learning*. Knowledgeable Object Symposium, Chau Chak Wing Museum.
FoxScan. (2025). Record fox activity in FoxScan. Accessed from https://www.feralscan.org.au/foxscan/map.aspx
Gibson, J. J. (1979). *'The Theory of Affordances'. The Ecological Approach to Visual Perception*. Boston, MA: Houghton Mifflin Harcourt (HMH), p. 127.
Gómez, E. (2013). Dog parks: Benefits, conflicts, and suggestions. *Journal of Park and Recreation Administration*, *31*(4), 79–91.
Gómez, E., Baur. J. W. R., & Malega, R. (2018). Dog park users: An examination of perceived social capital and perceived neighborhood social cohesion. *Journal of Urban Affairs*, *40*(3), 349–369. https://doi.org/10.1080/07352166.2017.1343634
Gorichanaz, T. (2020). *Information Experience in Theory and Design*. Bingley, UK: Emerald Group Publishing.
Gorichanaz, T. (2026). Information experience for the expanding circle: Insights from animal–computer interaction and more-than-human–computer interaction. In N. Solhjoo (Ed.), *Multispecies Information Science* (pp. 40–52). Routledge.
Graham, T. M., & Glover, T. D. (2014). On the fence: Dog parks in the (Un)leashing of community and social capital. *Leisure Science*, *36*, 217–234.
Granovetter, M. S. (1973). The strength of weak ties. *American Journal of Sociology*, *78*(6), 1360–1380.
Greater Sydney Parklands. (2025). Callan Park environment. *Greater Sydney Parklands; NSW Government*. Accessed from https://www.greatersydneyparklands.nsw.gov.au/learn-and-discover/environment-and-nature/callan-park-environment

Grenersen, G. (2016). Documentation, information and the animal connection. *Proceedings from the Document Academy*, *3*(2), 15. https://doi.org/10.35492/docam/3/2/15

Harris, L. D., & Wasilewski, J. (2004). Indigeneity, an alternative worldview: four R's (relationship, responsibility, reciprocity, redistribution) vs. two P's (power and profit). Sharing the journey toward conscious evolution. *Systems Research and Behavioral Science*, *21*(5), 489–503.

Harrison, S., & Dourish, P. (1996). Re-place-ing space: The roles of place and space in collaborative systems. In *Proceedings of the 1996 ACM Conference on Computer Supported Cooperative Work (CSCW)*, Boston, MA, 16–20 November, pp. 67–76.

Hodgetts, T. (2017). How we nose. In M. Bastian, O. Jones, N. Moore, & E. Roe (Eds.), *Participatory Research in More-Than-Human Worlds* (pp. 93–105). London: Routledge.

Hull, G. (2018). *The Dogs that Made Australia*. AU: Harper Collins.

Iacono, M. (2024). More Paws than Prams. *Darling Magazine*, Balmain Rozelle Chamber of Commerce. Accessed from https://www.balmainrozelle.org.au/darling/stories/more-paws-than-prams.html

Ingold, T. (2011). Against space: Place, movement, knowledge. In P. W. Kirby(Ed.), *Boundless Worlds* (1st ed., pp. 29–44). New York: Berghahn Books.

Irvine, L., & Cilia, L. (2017). More-than-human families: Pets, people, and practices in multispecies households. *Sociology Compass*, *11*(2), e12455. https://doi.org/10.1111/soc4.12455

Jackson, P. (2012). Situated activities in a dog park: Identity and conflict in human-animal space. *Society & Animals*, *20*(3), 254–272.

Jones, P., & Evans, J. (2012). Rescue geography: Place making, affect and regeneration. *Urban Studies*, *49*(11), 2315–2330.

Lloyd, A. (2006). Information literacy landscapes: An emerging picture. *Journal of Documentation*, *62*(5), 570–583.

Massey, D. (2013). *Space, Place and Gender*. New Jersey: John Wiley & Sons.

Masson, J. M. (1998). *Dogs Never Lie about Love: Reflections on the Emotional World of Dogs*. New York: Crown Publishing Group.

Miller, G. A. (1993). Psychology and information. *ACM SIGDOC Asterisk Journal of Computer Documentation*, *17*(3), 3–6. https://doi.org/10.1145/165552.165556

Montague, K. E. (2024). Mapping the road ahead: understanding social factors that shape vehicle residents' information grounds. *Information Research an International Electronic Journal*, *29*(2), 427–435. https://doi.org/10.47989/ir292838

Narayan, B. (2013). From everyday information behaviours to clickable solidarity in a place called social media. *Cosmopolitan Civil Societies: An Interdisciplinary Journal*, *5*(3), 32–53.

Nardi, B. A., & O'Day, V. (1999). *Information Ecologies: Using Technology with Heart*. Cambridge, MA: MIT Press.

NSW Government. (2002). Callan Park (Special Provisions) Act 2002 No 139. *NSW Legislation; NSW Government*. Accessed from https://legislation.nsw.gov.au/view/html/inforce/current/act-2002-139

O'Hare, P. (2024). A walk with "that wild dog of yours": tales of circumscribed, co-negotiated and adaptive walking practices. *Social & Cultural Geography*, *25*(8), 1311–1330.

Oldenburg, R. (1999). *The Great Good Place: Cafes, Coffee Shops, Bookstores, Bars, Hair Salons, and other Hangouts at the Heart of a Community*. New York: Da Capo Press.

Orlean, S (2011). The social animal: The rules of dog park etiquette. *The New Yorker*, October 11, 2011. Accessed from https://www.newyorker.com/culture/susan-orlean/the-social-animal 1/7

Pemberton, N. (2017) The Burnley Dog War: The politics of dog-walking and the battle over public parks in post-industrial Britain. *Twentieth Century British History*, 28(2), 239–267. https://doi.org/10.1093/tcbh/hww045

Pettigrew, K. E. (1999). Waiting for chiropody: Contextual results from an ethnographic study of the information behaviour among attendees at community clinics. *Information Processing & Management*, 35(6), 801–817.

Pink, S. (2015). *Doing Sensory Ethnography*. London: Sage.

Proctor, C. (2017). *Sydney for Dogs*. Warriewood, NSW: Woodslane Press.

Robins, D. M., Sanders, C. R., & Cahill, S. E. (1991). Dogs and their people: Pet-facilitated interaction in a public setting. *Journal of Contemporary Ethnography*, *20*(1), 3–25.

Rowley, J. (1998). What is information? *Information Services and Use*, *18*(4), 243–254.

Rutherford, D. (2020). Funding anthropological research in the age of COVID-19. In V. Das & N. Khan(Eds.), *Covid-19 and Student Focused Concerns: Threats and Possibilities*. American Ethnologist website, May 1, 2020. Accessed from https://americanethnologist.org/features/collections/covid-19-and-student-focused-concerns-threats-and-possibilities/funding-anthropological-research-in-the-age-of-covid-19

Savolainen, R. (2007). Information behavior and information practice: reviewing the "umbrella concepts" of information-seeking studies. *The Library Quarterly*, *77*(2), 109–132.

Savolainen, R. (2009). Small world and information grounds as contexts of information seeking and sharing. *Library & Information Science Research*, *31*(1), 38–45.

Shamayleh, G., & Arsel, Z. (2025). Digital affective encounters: The relational role of content circulation on social media. *Journal of Consumer Research*, ucaf023. https://doi.org/10.1093/jcr/ucaf023

Solhjoo, N. (2024a). The red thread of information across more-than-human love. *Journal of Librarianship and Information Science*. https://doi.org/10.1177/09610006241255418

Solhjoo, N. (2024b). Knowing within multispecies families: An information experience study. *Journal of Information Science*. https://doi.org/10.1177/01655515241268845

Solhjoo, N. (2024c). Information experience within multispecies families: Understanding shared life with companion cats and dogs (Doctoral dissertation, Te Herenga Waka—Victoria University of Wellington). Victoria University of Wellington Open Access Repository. Accessed from https://openaccess.wgtn.ac.nz/articles/thesis/Information_Experience_within_Multispecies_Families_Understanding_shared_life_with_companion_cats_and_dogs/27236811

Solhjoo, N., Krtalić, M., & Goulding, A. (2024). The in-between: Information experience within human-companion animal living environments. *Journal of Documentation*, *80*(4), 701–729. https://doi.org/10.1108/JD-08-2023-0160

Sonnenwald, D. H. (1999). *Evolving Perspectives of Human Information Behavior: Contexts, Situations, Social Networks and Information Horizons*. London: Taylor Graham.

Springgay, S., & Truman, S. (2018). *Walking Methodologies in a More-than-Human World: Walking Lab*. Published: New York: Routledge.

Talip, B. A., Narayan, B., Edwards, S. L., & Watson, J. (2016). Digital ethnography as a way to explore information grounds on Twitter. *Qualitative and Quantitative Methods in Libraries*, *5*(1), 89–105.

Thomas, E. M. (2010). *The Hidden Life of Dogs*. Boston, MA: Houghton Mifflin Company, p. 283.

Tissot, S. (2011). Of dogs and men: The making of spatial boundaries in a gentrifying neighborhood. *City & Community*, *10*(3), 265–284.

Tsing, A. L. (2015). *The mushroom at the end of the world: On the possibility of life in capitalist ruins*. Princeton, NJ: Princeton University Press.
Tuan, Y. F. (1977). *Space and Place: The Perspective of Experience*. Minneapolis, MN: University of Minnesota Press.
Tuan, Y. F. (1984). *Dominance and Affection: The Making of Pets*. New Haven, CT: Yale University Press.
Urbanik, J., & Morgan, M. (2013). A tale of tails: The place of dog parks in the urban imaginary. *Geoforum*, *44*, 292–302.
Vanni, I., & Crosby, A. (2023). Place-based methodologies for design research: An ethnographic approach. *Design Studies*, *85*, 101168.
Vincent, A. (2019). Dog parks as an institutional resource for social capital in the urban neighborhood. *Society & Animals*, *29*(5–6), 517–538.
Uexküll, J. von (2013). *A Foray Into the Worlds of Animals and Humans: With a Theory of Meaning* (Vol. 12). Minneapolis, MN: Universityof Minnesota Press.
Williamson, K. (1998). Discovered by chance: the role of incidental information acquisition in an ecological model of information use. *Library and Information Science Research*, *20*(1), 23–40. https://doi.org/10.1016/S0740-8188(98)90004-4
Wise, A. (2008). "It's just an attitude that you feel": Interethnic habitus before the Cronulla riots. In G. Noble (Ed.), *Lines in the Sand: The Cronulla Riots and the Limits of Australian Multiculturalism* (pp. 127–145). Sydney, Australia: Institute of Criminology Press.
Wood, L., & Christian, H. (2011). Dog walking as a catalyst for strengthening the social fabric of the community. In R. Johnson, A. Beck, & S. McCune (Eds.), *The Health Benefits of Dog Walking for People and Pets: Evidence and Case Studies* (pp. 51–74). Indiana: Purdue University Press.
Xuanyu, Y., & Xiaoyu, Z. (2023). A multi-species ethnographic perspective on urban human-wildlife relationships—A comparative study of stray cats and wild raccoon dogs in Shanghai city. *Academic Journal of Humanities & Social Sciences*, 6(4), 73–78. https://doi.org/10.25236/AJHSS.2023.060414

10

FROSTED MUZZLES AND DIGITAL BONDS

Embodied Information Practices and Affective Adoption in a Crisis Shelter Ecology

Monika Krakowska and Magdalena Zych

Introduction

The winter of 2024 in Poland brought both a sharp temperature drop and a surge in digital mobilisation around animal welfare. Amidst plummeting temperatures, a crisis unfolded in local dog shelters, where dozens of animals, both in and outside shelters, faced the risk of freezing to death. In response, a grassroots initiative, informally dubbed the 'Frost Action,' emerged, led by a non-profit organisation that sought to rescue, rehome, and sustain the lives of abandoned dogs through emergency adoptions. The episode's significance lies not only in its emotional appeal or public reach but in how information operated within this multispecies emergency—shaped by embodied and affective infrastructures (Solhjoo, 2024a).

The chapter explores how, during the crisis, adopters formed ethical commitments to dogs based on minimal yet emotionally charged information: Photos, symbolic captions, and brief narratives (e.g., 'senior,' 'cold,' 'gentle'). Often adopting without prior contact or verification, they constructed imagined interspecies bonds through affective projections under conditions of informational uncertainty. Many respondents described being moved by imagined suffering and the projection of intimacy towards an absent other. These micro-narratives gave rise to an epistemic situation in which affect became the primary organising principle of information behaviour, generating what might be called embodied imaginaries of multispecies care, constructed through somatic metaphor, emotional projection, and embodied simulation (Lakoff & Johnson, 1999; Barsalou, 2008). Grounded in multispecies information science and informed by theories of embodied cognition (Bates, 2006; Frohmann, 2009), crisis informatics, and epistemic trust (Fricker, 2007; Brożek, 2019),

DOI: 10.4324/9781003583424-12

this chapter also draws on recent studies that foreground affective relationality in human–nonhuman information experience (Gorichanaz, 2020; Solhjoo, 2024a; Solhjoo et al., 2024; Solhjoo, 2025). It addresses the following research question: How do humans construct embodied and affective information practices when making irreversible decisions about nonhuman others under conditions of absence, urgency, and mediated trust?

The Frost Adoption Action (2024)

Drawing on a qualitative case study, the Frost Action campaign was analysed as a multispecies intervention shaped by affective and epistemic artefacts. The campaign relied on curated Facebook content—photographs, symbolic captions (e.g., *senior*, *cold*, *invisible*), and condensed narrative appeals—that transformed complex realities into emotionally salient cues. These fragments circulated widely, generating hundreds of comments that blurred the boundary between observation, emotional projection, and ethical intent. Typical expressions included: 'Since I first saw her in that December post, I kept my fingers crossed for her...,' 'It was impossible to part with him,' or 'He's so focused on his new guardian; it doesn't surprise us at all.' Such utterances reveal how emotional attachment and symbolic framing co-constituted the campaign's informational environment, entangling human and animal subjectivities in imagined mutuality.

Building on Krakowska and Zych (2025), this chapter shifts the focus from human information flow to interspecies relationality, affective infrastructures, and the performative role of information in ethically charged encounters. It engages with relational epistemology, multispecies agency, and affective ethics, arguing that in multispecies crises, informational agency involves not just knowledge-seeking but also emotional projection and embodied connection, often through imagined tactile intimacy or 'touch talk' (Solhjoo, 2024a), forming intersubjective bonds across species.

Beyond informing, these materials evoked emotional and moral responses. Close-up photography, emotionally charged hashtags, and semiotic condensation transformed the animal's face into an interface for multispecies meaning-making. These artefacts may be understood as documents in motion (Frohmann, 2009) or as *multispecies signals*—affect-saturated fragments that catalyse ethical response and enact what may be called informational kinship: A form of epistemic and ethical alignment enacted through shared vulnerability, somatic resonance, and projected care.

This study uses publicly available digital materials such as Facebook posts, images, and narratives. Guided by the 'information as experience' paradigm (Gorichanaz, 2020), it emphasises affective and relational meaning-making. Methodological details are in Krakowska and Zych (2025). From an epistemic justice perspective (Fricker, 2007), the campaign shows interspecies testimonial injustice. Dogs were made epistemically mute, their subjectivity inferred through human emotion. This asymmetry reveals the challenge of interpreting nonhuman signals without anthropocentric bias, though such projections proved ethically powerful.

Emotional inference, often seen as subjective, should be valued as valid information processing in multispecies and crisis contexts. Information is evoked, embodied, and enacted, calling for a move beyond human-centric models towards relational and situated frameworks.

Theoretical Background

This chapter draws on multispecies information science, crisis informatics, and embodied cognition to examine information behaviour during the Frost Action. It focuses on interspecies communication, the dog as a performative document, and knowledge-making amid emotional and high-entropy conditions.

Multispecies Information Practices

Recent work in information science increasingly recognises the epistemic, affective, and material entanglements between human and nonhuman agents. Scholars such as Bates (2026) and Solhjoo (2024a, 2024b, 2025) foreground relational infrastructures of meaning and care in interspecies contexts. Building on feminist technoscience (Haraway, 2023), multispecies ethnography (Kirksey & Helmreich, 2010), and relational epistemology (Fricker, 2007; Brożek, 2019), multispecies information science challenges anthropocentric assumptions in mainstream information behaviour research. Here, information emerges through embodied, affective encounters; meaning arises via gestures, odours, and representations within species-specific umwelten (Von Uexküll, 1957) (see also Marcia Bates's (2026) discussion in Chapter 2 of this volume, which draws on Jakob von Uexküll's idea of species-specific perceptual worlds). This aligns with Bates's (2006) view of information as organised matter and energy and with Worrall et al. (2021) and Krakowska (2020) on affect in ethically complex situations. Animals function as informational agents, shaping human ethical cognition and mobilising affective interpretation (cf. Haraway, 2003). These interspecies encounters generate Haraway's 'response-ability': A shared epistemic space where meaning is co-constituted through presence, perception, and attunement. In crisis, this becomes relational sense-making rooted in affective proximity and ethical imagination.

During the Frost Action, adopters responded to emotional cues that evoked care through minimal, morally potent fragments (Krakowska & Zych, 2025). These reflect Solhjoo's (2024a) affective infrastructures, configurations that prompt action without full analysis. Rather than aiding cognition, they reshaped epistemic engagement. Captions and hashtags served as effective triggers. For many adopters, these signs were enough. They aligned with cultural imaginaries of abandoned dogs and needed no further explanation. Some accepted the situation's opacity, assuming the dog would adapt or that care required no justification. Emotional resonance often replaced epistemic completeness. We introduce the term 'symbolic overload' to describe these affectively saturated signs that catalyse action under conditions of crisis (cf. Krakowska & Zych, 2025). These cues condensed ethical meaning into emotionally charged appeals to care. The dog becomes a relational co-agent, not a passive object. Adoption becomes an embodied, performative way of knowing that evokes multispecies futures rooted in care and responsibility.

The Dog as Embodied Document

Understanding the epistemic dynamics of Frost Action requires viewing the document as a relational, multisensory construct rather than a fixed symbolic artefact. While classical theory emphasises symbolic stability, Frohmann (2009) and Bates (2006) show that documents are performative and shaped by social, material, and affective contexts. Information is not only cognitive or representational but also embodied and relational, emerging through sensory experience and emotional attunement. In multispecies settings, it appears through postures, gestures, vocalisations, and cues that gain meaning through interaction. This aligns with Bates (2006) and Hartel (2010), who highlight the role of embodiment and affect in information.

The dog-as-document merges living subject and symbolic construct, enabling co-created meaning (cf. Frohmann, 2009; Brożek, 2019). This aligns with Hartel's (2005, 2010) concept of aesthetic information, where affect and sensation generate understanding. Documents here are not only read but felt, inviting ethical engagement rather than delivering full knowledge. This approach aligns with recent efforts to expand documentary theory to include affective, relational, and multisensory forms (Gorichanaz, 2020). In Frost Action, adopters responded not to empirical data but to emotionally compressed representations. This reflects Solhjoo's (2024a) concept of symbolic compression, where semiotic fragments condense ethical meaning and prompt action. The dogs, as living documents, embodied vulnerability and the possibility of interspecies intimacy, dissolving boundaries between artefact and subject, knowledge and care.

Information Complexity in Crisis

The Frost Action reveals an epistemic condition where knowing, feeling, and acting converge. Information is activated through care, urgency, and ethical projection. As crisis informatics research shows (Palen & Anderson, 2016; Krakowska, 2020), decisions made under emotional and ethical pressure reflect altered epistemic conditions. Instead of information scarcity, individuals face a symbolic surplus—fragmented images, emotive cues, and condensed narratives that require swift interpretation and action. There is also the notion of epistemic triggers (Krakowska, 2020), affect-laden cues that prompt immediate action rather than deliberation. Complexity arises not from informational ambiguity but from symbolic overload and interpretive underdetermination. Following Savolainen's (2017) theory of everyday life information seeking, such actions unfold within emotionally framed horizons of relevance and social norms. Starbird et al. (2014) show that in digital crises, people rely on distributed sense-making, where emotional resonance and redundancy replace traditional verification. These frameworks help explain how Frost Action adopters navigated unstable informational terrain through projected relationality rather than factual triangulation.

Epistemic Trust and Affective Closure

The Frost Action adoption process illustrates a mode of epistemic engagement where decisions were made without direct contact, verification, or full knowledge. Traditional models assume a deliberative agent who seeks and verifies to reduce uncertainty, but under constraint, affect, and ethical urgency, these models become inadequate. Here, epistemic trust is a primary orientation to action (Fricker, 2007; Jäger, 2016). This trust arose from moral alignment, emotional resonance, and care. Nonprofits, volunteers, and adopters acted as situated epistemic authorities (Zagzebski, 2012), gaining credibility through affective legitimacy. Visual cues served as epistemic triggers (Krakowska, 2020), prompting moral responses without requiring full information. These practices formed epistemic infrastructures (Jäger, 2016), affective networks where trust, not evidence, enabled knowing and action.

Central to this is affective closure, when emotional certainty replaces epistemic completeness. Cognitive resolution gives way to ethical sufficiency, with epistemic closure (Brożek, 2019) reframed as resonance-based necessity. Crucially, affective trust is not a failure of reason but a relational mode of knowing, suited to conditions of urgency and asymmetry.

Fricker's (2007) concept of epistemic injustice takes on a multispecies dimension: Animals, though not human 'knowers,' become epistemic referents anchoring meaning and moral action. Frost Action reveals a shift in crisis information behaviour, where trust forms an infrastructure of care. Knowing merges with feeling, and action is justified through affective attunement rather than verification.

Findings and Interpretative Threads

Affective Filtering and Recognition

Frost Action participants engaged with information through emotional cues signalling urgency and moral relevance. Images of shivering dogs and captions like *'waiting for warmth'* served as affective filters, drawing attention and prompting immediate ethical responses. Users often interpreted dogs' gestures through emotional familiarity: An affective resonance that bypassed reflection. This aligns with Krakowska's (2020, 2022) concept of epistemic triggers: Emotionally charged cues that provoke action through symbolic intensity and moral urgency under uncertainty. Often, such triggers did not generate new representations but resonated with existing mental images of suffering. They functioned as *epistemic closures*, completing partial or dormant schemas and rendering cues like *'cold,' 'urgent,'* or *'senior'* subjectively sufficient. Action followed not despite uncertainty but because emotional saturation made further inquiry feel unnecessary. For some adopters, the belief that *'anyone can care for a dog'* effectively foreclosed deeper verification. Others consciously accepted ambiguity, trusting that any behavioural unpredictability would resolve within a new relational context.

Selective perception served as an affective-epistemic strategy, grounding meaning in emotional salience. Images of touch, gaze, or closeness between dogs and humans enabled recognition and enacted care. Participants reported visceral responses rooted in vulnerability and ethical immediacy rather than analysis.

Information Domestication and Closure

There is no single, stable image of a dog; adoptability is shaped through affective cues, linguistic framings, and situational assumptions (Nakamura et al., 2019). This applies to both language and visuals. Descriptors such as *'lively'* or *'independent'* have been linked to shorter shelter stays, while terms like *'only dog'* often correlate with longer ones, suggesting that profiles construct, rather than merely describe, a dog's persona. Likewise, visual cues such as facial expression, eye contact, or the presence of a barred enclosure influence perceived sociability and adoption appeal (Lamb, Andrukonis, & Protopopova, 2021).

After initial recognition, many participants described a sense-making process that transformed the dog from a profile into an imagined companion. They recalled vivid images of the dog in their home, such as lying on a couch, under a blanket, or *'finally warm and loved.'* These projections were emotionally grounded. Few mentioned verifying details or seeking more context. Instead, they described a form of affective certainty: *'I just knew,' 'it was a sign,'* or *'something touched me and I couldn't ignore it.'* This form of closure did not rest on factual completeness but on emotional sufficiency. Rather than constructing a comprehensive profile, individuals appeared to reach a threshold where available cues were enough to justify action. For many, this culminated in an irreversible step: Submitting an adoption request, coordinating transport, or widely sharing the post. At that moment, the dog ceased to be hypothetical; it became part of a projected relational future.

Multispecies Agency and Meaning-Making

Adopters described the dog not as a passive recipient of care but as an active participant in the encounter. Users frequently portrayed the animal as reaching out: '*Choosing,*' '*calling,*' or '*asking*' for rescue. Comments such as '*Those lovely, faithful doggy eyes*' or '*How much happiness on those faces!*' illustrate how emotional and symbolic projections imbued the dog with agency. In these accounts, the dog was a co-agent in a relational process. This co-agency was often communicated through perceived gestures: Sustained gaze, expressive posture, or a look that '*went straight to the heart.*' These perceptions were accompanied by somatic responses: '*goosebumps,*' '*tears,*' or '*a tightness in the chest,*' which participants described as overwhelming and impossible to ignore. Embodied reactions were not limited to physical encounters. Even mediated through photographs or short videos, dogs were described as '*present,*' '*visible in their soul,*' or '*reaching out from the screen.*' For many, this sense of presence prompted immediate action: Adopting, donating or sharing the post. These relational perceptions formed a felt reality in which the relationship was already unfolding. Participants responded not to symbolic figures but to animals sensed as already engaged in mutual recognition.

Information practices in Frost Action were based not on verification but on embodied attunement and ethical immediacy. Participants did not confirm decisions but enacted them, responding to emotionally potent cues as invitations to care. Information was not validated but lived, co-produced through multispecies presence, affect, and symbolic condensation that enabled irreversible action.

Ethical Imaginaries and Embodied Decisions

Adoption during Frost Action relied not on rational choice but on ethical imaginaries, affective-cognitive frameworks that supported action under uncertainty by envisioning care and multispecies belonging. These emerged through sensory and symbolic cues like images of shivering dogs, captions of abandonment, and labels such as 'senior' or 'cold.' The decision wasn't made about the dog but with the dog, through minimal cues that fostered recognition. These imaginaries were shaped by the campaign's socio-technical infrastructure, which turned complexity into moral clarity.

This communicative ensemble acted as epistemic scaffolds, enabling action without full information. Brożek's (2019) *epistemic closure* explains how people act under uncertainty through *coherent enough* internal worlds. In Frost Action, closure came through affect, not analysis. Participants spoke of bodily knowing ('I just knew,' 'her eyes spoke to me,' 'it felt like he chose me'), echoing Ahmed's (2004) *affective orientations* that guide action beyond reasoning.

The animal, though silent, was not passive but a co-constituent in the epistemic encounter, a living document (Bates, 2006; Frohmann, 2009) whose meaning emerged through experience rather than data. Hartel (2010) shows that in domains like gourmet cooking, affect and sensory engagement can override analysis. Similarly, the dog was encountered as a presence, situated within imagined futures of rescue and companionship. Frost Action shows how ethical imaginaries act as affective infrastructures of decision-making, allowing individuals to turn fragments into irreversible acts of care. These imaginaries are not epistemic failures but forms of moral world-building, where information, affect, and urgency are inseparable.

Discussion

A primary theoretical contribution of this study lies in the redefinition of document theory through embodied and relational epistemologies. Frost Action demonstrates that document needs may emerge dynamically through affective projection, somatic resonance, and symbolic condensation. The image or name of a dog encountered via social media functioned as a *document in motion* (Frohmann, 2009): A conduit of ethical meaning shaped by affective urgency and imaginative projection. Drawing on Bates's (2006) call to understand documents within lived practice, this study extends the notion into a multispecies domain. Dogs, as embodied presences within affective infrastructures, functioned as *relational documents* encountered cognitively, somatically, and ethically. This process exemplifies Brożek's (2019) model of epistemic closure through interpretive completion: The idea that individuals act not when full information is available, but when internal coherence suffices. In Frost Action, closure did not emerge from deliberation or verification but from emotional sufficiency, described by adopters as 'a call of the heart,' 'something that touched me,' or 'a sign.' These affective confirmations catalysed decisive action and often transformed lives. Closure was enacted through somatic alignment, not rational analysis.

The notion of information domestication emerged as participants imagined the dog within their home, reframing it as a projected companion. Within information experience theory (Bruce et al., 2014; Hartel, 2020), this shows how minimal cues can evoke affective simulations of multispecies futures. Drawing on Barsalou's (2008) theory of situated conceptualisation, emotionally salient fragments like an image or caption enabled adopters to pre-enact care and belonging.

Another key insight concerns multispecies agency. Participants saw dogs not as passive recipients but as co-actors in the informational encounter. Phrases like 'he chose me' or 'she's asking for love' show how agency was co-constructed across species. These moments involved embodied responses such as goosebumps, tears, or chest tightness, interpreted as signs of mutual recognition. Though animals did not speak, they were perceived as communicating effectively, prompting epistemic and moral commitments. This aligns with relational epistemologies, where knowledge is situated, embodied, and co-produced in affective contexts (Fricker, 2007; Shah et al., 2023).

This recognition prompts reflection on testimonial injustice in multispecies contexts (Fricker, 2007). Dogs' subjectivity was often framed through anthropocentric projections of emotion, such as sadness or gratitude. Though epistemically thin, these projections were ethically thick, enabling care and prompting action. Rather than dismissing them as unreliable, this chapter supports a reflexive multispecies methodology that values affective inference in urgent contexts while remaining aware of the limits of anthropomorphism (see Solhjoo, 2026, Chapter 7).

Frost Action challenges conventional models of trust in information behaviour. Trust was based not on institutional authority or evidence but on embodied attunement, a felt alignment between emotional cues and moral urgency. This reflects Papacharissi's (2016) affective publics and Bosworth's affective infrastructures, where symbolic content circulates in emotionally charged interpretive fields. Epistemic authority is ambient, distributed, and affectively driven. The dog's image on social media did not function as a static representation but as a catalyst for shared orientation.

These affective practices formed epistemic infrastructures (cf. Jäger, 2016; Tichenor et al., 2022): Networks of credibility, emotion, and symbolism that supported ethical decisions without full information. They enabled users to act on felt coherence, grounded in care and somatic resonance rather than certainty or analysis.

This chapter calls for recognising documentality and practice must include nonhuman actors as co-constitutive participants in meaning-making. Day's (2026) critique of traditional documentation in Chapter 4, where he argues for a new understanding of documentality—one in which meaning emerges from an entity's own affordances and its ability to interact and express itself within assemblages. Frost Action shows how affective presence, somatic resonance, and ethical imagination shape information practices across species. Information acted as a *medium of ethical orientation*, enabling new futures to be imagined, inhabited, and enacted in advance. This logic of informational kinship invites a revision of how we understand knowledge, communication, and agency in multispecies entanglements.

Conclusions

Frost Action shows that in crisis, information functions as an embodied, affective, and multispecies force. Adopters engaged not analytically but through *symbolic overload*, responding to emotionally charged cues that prompted ethical action. Decisions arose from relational projection and moral urgency rather than verification. Information was not simply received but enacted as embodied care amid uncertainty.

This case prompts a rethinking of document theory and information agency. The dogs acted as *living documents*: Relational, embodied carriers of meaning that spurred affective commitment. These documents emerged through *symbolic cues, somatic resonance*, and *affective inference*. In urgent, interspecies contexts, documentation is *performative*, co-produced through mediated presence and cross-species attunement. Their authority arose not from institutional credibility, but from *emotional salience* and anticipatory care. Findings support a broader turn in information science towards *embodied, experiential*, and *multispecies* perspectives on information behaviour. In *Frost Action*, affective infrastructures enabled meaningful decisions under conditions of ambiguity, distilling informational complexity into minimal but symbolically saturated cues. This process produced a *threshold of affective certainty*: A state in which ethical action became justified not by evidence, but by resonance. Future research should further explore how *affective publics* and *epistemic infrastructures* (Jäger, 2016; Papacharissi, 2016) shape collective responses in crisis, how adoption narratives mobilise *ethical imaginaries*, and how *multispecies co-presence* transforms the movement and meaning of information across digital platforms. Studying such dynamics demands frameworks that attend not only to what information *is* but also to what it *does*: How it traverses bodies, reconfigures relationships, and enables ethical world-making.

As this chapter has shown, information in crisis is not neutral. It becomes a conduit of urgency, a medium of relational projection, and a catalyst of transformation across species, platforms, and lives. For many adopters, action was based not on verification but on felt certainty: An emotional pull grounded in care, imagined futures, and the belief that 'she wasn't just a picture, she was calling me.' This may be the heart of multispecies information: Not evidence but a moment of recognition—where love becomes a way of knowing.

References

Ahmed, S. (2004). *The Cultural Politics of Emotion*. Routledge. https://doi.org/10.4324/9780203700372

Barsalou, L. W. (2008). Grounded cognition. *Annual Review of Psychology*, 59, 617–645. https://doi.org/10.1146/annurev.psych.59.103006.093639

Bates, M. J. (2006). Fundamental forms of information. *Journal of the American Society for Information Science and Technology*, 57(8), 1033–1045. https://doi.org/10.1002/asi.20369

Bates, M. J. (2026). Researching information behaviour among multiple species. In N. Solhjoo (Ed.), *Multispecies Information Science* (pp. 23–28). Routledge.

Bosworth, K. (2022). What is 'affective infrastructure'? *Dialogues in Human Geography*, 13(1), 54–72. https://doi.org/10.1177/20438206221107025

Brożek, B. (2019). *Granice interpretacji*. Copernicus Center Press.

Bruce, C., Davis, K., Hughes, H., Partridge, H., & Stoodley, I. (2014). Information experience: Contemporary perspectives. In C. Bruce, K. Davis, H. Hughes & H. Partridge (Eds.), *Information Experience: Approaches to Theory and Practice* (pp. 3–15). Emerald Group Publishing Limited. https://doi.org/10.1108/S1876-056220140000010001

Day, R. E. (2026). The animal's importance for neo-documentation: A "copernican revolution" in understanding information. In N. Solhjoo (Ed.), *Multispecies Information Science* (pp. 53–60). Routledge.

Fricker, M. (2007). *Epistemic Injustice: Power and the Ethics of Knowing*. Oxford University Press. https://doi.org/10.1093/acprof:oso/9780198237907.001.0001

Frohmann, B. (2009). Revisiting what is a document? *Journal of Documentation*, 65(2), 291–303. https://doi.org/10.1108/00220410910937624

Gorichanaz, T. (2020). *Information Experience in Theory and Design*. Emerald Publishing Limited.

Haraway, D. (2003). When species meet. In A. Franklin (Ed.), *The Routledge International Handbook of More-than-Human Studies* (pp. 42–78). Routledge. https://doi.org/10.4324/9781003262619

Hartel, J. (2005). Serious leisure. In K. Fisher, S. Erdelez, & L. McKechnie (Eds.), *Theories of Information Behavior: A Researcher's Guide* (pp. 313–317). Information Today.

Hartel, J. (2010). Time as a framework for information science: Insights from the hobby of gourmet cooking. *Information Research*, 15(4), 14–15. https://informationr.net/ir/15-4/colis715.html

Hartel, J. (2020). The red thread of information. *Journal of Documentation*, 76(3), 647–656. https://doi.org/10.1108/JD-04-2019-0067

Jäger, C. (2016). Epistemic authority, preemptive reasons, and understanding. *Episteme*, 13(2), 167–185.

Kirksey, S. E., & Helmreich, S. (2010). The emergence of multispecies ethnography. *Cultural Anthropology*, 25(4), 545–576. https://doi.org/10.1111/j.1548-1360.2010.01069.x

Krakowska, M. (2020). Affective factors in human information behavior: A conceptual analysis of interdisciplinary research on information behavior. *Zagadnienia Informacji Naukowej–Studia Informacyjne*, 58(1A[115A]), 75–95.

Krakowska, M. (2022). *Zachowania informacyjne człowieka w kontekście zjawiska epistemicznej bańki informacyjnej : propozycja nowej koncepcji*. [*Human information behavior in the context of the epistemic information bubble phenomenon: A proposal for a new concept*]. Wydawnictwo Uniwersytetu Jagiellońskiego w Krakowie.

Krakowska, M., & Zych, M. (2025). Social responsibility of NGOs in the shaping of information behaviour in emergency situations – A case study of the frost action. *Journal of Documentation*, ahead-of-print. https://doi.org/10.1108/JD-01-2025-0030

Lakoff, G., & Johnson, M. (1999). *Philosophy in the Flesh: The Embodied Mind and its Challenge to Western Thought*. Basic Books.

Lamb, F., Andrukonis, A., & Protopopova, A. (2021). The role of artificial photo backgrounds of shelter dogs on pet profile clicking and the perception of sociability. *PLOS ONE*, 16(12), e0255551. https://doi.org/10.1371/journal.pone.0255551

Nakamura, M., Dhand, N. K., Starling, M. J., & McGreevy, P. D. (2019). Descriptive texts in dog profiles associated with length of stay via an online rescue network. *Animals*, 9(7), 464. https://doi.org/10.3390/ani9070464

Palen, L., & Anderson, K. M. (2016). Crisis informatics—New data for extraordinary times. *Science*, 353, 224–225. https://doi.org/10.1126/science.aag2579

Papacharissi, Z. (2016). Affective publics and structures of storytelling: Sentiment, events and mediality. *Information, Communication & Society*, 19(3), 307–324. https://doi.org/10.1080/1369118X.2015.1109697

Savolainen, R. (2017). Everyday life information seeking. In J. D. McDonald, & M. Levine-Clark (Eds.), *Encyclopedia of Library and Information Sciences: Fourth Edition* (pp. 1506–1515). CRC Press. https://doi.org/10.1081/E-ELIS4

Shah, R., Ahluwalia, S., & Spicer, J. (2023). Relational care and epistemic injustice. *Primary Health Care Research & Development*, 24, e62. https://doi.org/10.1017/S1463423623000555

Solhjoo, N. (2024a). The red thread of information across more-than-human love. *Journal of Librarianship and Information Science*. https://doi.org/10.1177/09610006241255418

Solhjoo, N. (2024b). Knowing within multispecies families: An information experience study. *Journal of Information Science*. https://doi.org/10.1177/01655515241268845

Solhjoo, N. (2025). The multispecies perspective in library and information science. *Information Research*, 30(COLIS), 323–330. https://doi.org/10.47989/ir30CoLIS51898

Solhjoo, N. (2026). Research pathways for multispecies information science. In N. Solhjoo (Ed.), *Multispecies Information Science* (pp. 91–110). Routledge.

Solhjoo, N., Krtalić, M., & Goulding, A. (2024). The in-between: Information experience within human-companion animal living environments. *Journal of Documentation*, 80(4), 701–729. https://doi.org/10.1108/JD-08-2023-0160

Starbird, K., Muzny, G., & Palen, L. (2014). Learning from the crowd: Collaborative filtering techniques for identifying on-the-ground Twitterers during mass disruptions. *Proceedings of the 11th International ISCRAM Conference*, 1–10. https://cmci.colorado.edu/~palen/Home/Articles_by_Year_files/LearningfromCrowdStarbirdMuznyPalen.pdf

Tichenor, M., Merry, SE, Grek, S., & Bandola-Gill, J. (2022). Global public policy in a quantified world: Sustainable development goals as epistemic infrastructures. *Policy and Society*, 41(4), 431–444. https://doi.org/10.1093/polsoc/puac015

Von Uexküll, J. (1957). A stroll through the worlds of animals and men: A picture book of invisible worlds. In C. H. Schiller (Ed.), *Instinctive Behavior: The Development of a Modern Concept* (pp. 5–80). International Universities Press.

Worrall, A., Cappello, A., & Osolen, R. (2021). The importance of socio-emotional considerations in online communities, social informatics, and information science. *Journal of the Association for Information Science and Technology*, 72(10), 1247–1260. https://doi.org/10.1002/asi.24489

Zagzebski, L. (2012). *Epistemic Authority: A Theory of Trust, Authority, and Autonomy in Belief*. Oxford University Press.

11

BUILDING A BONSAI BOND

Pursuit of Perfection through a Person–Plant Partnership

Yazdan Mansourian

Introduction

Have you ever reflected on the bonds between human beings and plants? Your answer might be yes or no. In either case, as you are reading this chapter, I invite you to join me and reflect on this issue. We just need a few moments of contemplation to remember that this relationship is so fundamental that our existence depends on it. First, consider the air we breathe and the food we eat. Plants produce oxygen through photosynthesis, which is essential for our survival. Also, all kinds of food, whether they are grown directly or produced indirectly by animals, ultimately have a plant base.

However, our connection with plants is far beyond the essentiality of air and food. Many of the clothes we wear, as well as various household items such as tables and chairs, are made from wood. Although wooden items were more common in the past, they still are important in modern times. There is also a strong emotional bond between us and plants. Many people have a special tree in an old garden that reminds them of their childhood memories, or a potted plant gifted by a friend with emotional value. These natural companions often have a deeper meaning than their utilitarian uses.

In this chapter, I explore one of the profound expressions of this connection. I reflect on one of my recent research projects about the relationship between bonsai enthusiasts and their miniature trees. Bonsai is much more than a form of gardening. It is an artistic endeavour with a philosophical foundation that requires skills, dedication and perseverance. Bonsai growers who emphasise artistic control may create a risk of treating the tree as a medium. However, they should prioritise the trees' wellbeing alongside their aesthetic

DOI: 10.4324/9781003583424-13

goals (Brelje, 2023). Those who practise this art nurture their trees like a companion or even a member of their family. They pay attention to the tree's biological needs and listen to the tree. This special connection creates a bond that goes beyond aesthetic aspects. They shape living works of art that reflect both beauty and their own personal journey. They achieve these goals through constant attention and loving care that I explored in my research.

My Research Narrative

I share my story of how I became interested in studying bonsai as a hobby. Without sharing this story, some important aspects of the topic would remain untold. My research journey on the concept of *serious leisure* began in 2017. Serious leisure includes hobbies, voluntary activities and amateurism. Based on a systematic literature review outlining trends in the field (Mansourian, 2020), I conducted an empirical study involving interviews with serious leisure enthusiasts across diverse hobbies such as pottery and cycling (Mansourian, 2021b, 2021c). In the middle of that project, I came across a great example of serious leisure when one of the interviewees shared her story as a bonsai grower. In that moment, I realised that bonsai growing perfectly meets all the criteria of serious leisure. Accordingly, I dedicated my next project entirely to bonsai as a case study (Mansourian, 2021a, 2022a).

Bonsai growing is a traditional Japanese art form that involves growing trees in pots or shallow containers to resemble natural landscapes in a miniature size (Norman, 2014; Prescott, 2015; Chan, 2018). The term 'bonsai' is the Japanese version of the ancient Chinese art called 'penjing' or 'penzai,' with two thousand years of history (Zhao, 2012). Bonsai is a hobby that needs skill and commitment, and at the same time, it creates personal and social rewards that align with the framework of serious leisure. It needs a deep understanding of horticulture and artistic principles. Bonsai enthusiasts must learn a lot about pruning techniques, wiring methods and soil composition. Mastery in this art requires several years of patience and practice. Therefore, it is not a fleeting pastime. Bonsai trees are delicate and require meticulous shaping and attention. That is why, in my interviews, bonsai growers told me that it is an easy hobby to begin but a serious one to continue.

Many bonsai growers see themselves as artists or custodians of living sculptures. Their identity becomes intertwined with their practice, and they often join bonsai clubs. Joining a club was one of the main points that my interviewees mentioned during the interviews as advice to the newcomers. In terms of benefits, engaging in bonsai art brings both tangible and intangible rewards. Growers experience personal satisfaction, a sense of accomplishment, and even a meditative connection with nature. Their trees can live for generations, creating a legacy that extends beyond their own lifetime. That is why some of them believed: *You never own a bonsai; the bonsai owns you*!

Of course, while many engage in bonsai as a hobby, there are professional pathways available. Some enthusiasts become bonsai artists, bonsai teachers or competition judges. Others write books, produce educational content, or participate in international exhibitions. Each of these elements confirms that bonsai growing is not just a simple pastime. It is a deeply engaging and meaningful form of serious leisure.

Also, bonsai art is an ideal choice for studying information behaviour because it is an information-intensive hobby. Learning bonsai techniques requires lots of information seeking and sharing. They need to understand horticultural science, artistic design principles, and plant care, all of which involve seeking, organising, sharing, processing, evaluating, and applying information over time. The complexity of the hobby means that bonsai growers often engage in different types of information seeking, from structured learning, such as attending courses, to informal knowledge sharing within communities such as their clubs and societies. On the other hand, bonsai growers produce a massive volume of information in various formats, including books, magazine articles, YouTube videos, workshops, etc.

Moreover, bonsai growing is an information-intensive hobby because each tree requires specific care based on its species, age, and environmental conditions. Growers must constantly monitor changes. They need to diagnose problems such as pests or diseases. They also should adjust their techniques accordingly. This process involves gathering, organising, sharing, and interpreting information. The collaborative nature of the bonsai community further enhances its information richness, as practitioners share ideas. The findings of that project have already been published (Mansourian, 2024) and are not repeated here. Instead, this chapter engages with multispecies information science, which decentralises human dominance in narratives of information research. Therefore, I combine my phenomenological reflection with an attention to the agency of bonsai itself, which is based on the multispecies approach in Chapter 7 (Solhjoo, 2026), to develop and address a new research question:

- How does the practice of bonsai growing embody mutual care and information exchange between humans and plants?

Mapping My Research into the Related Scholarship

The *Serious Leisure Perspective* (SLP) is the main theoretical framework that I have been using in my ongoing research programme from the beginning. The term *serious leisure* was coined by Robert Stebbins in 1982, and he has been a prolific author on this topic ever since (Stebbins, 1982, 1992, 1997, 2001, 2007, 2020, 2021). A leisure pursuit is considered *serious leisure* if it has six qualities. First, those involved must be passionate about their chosen activity. Therefore, they commit time and effort over a long term as a hobby or a voluntary activity. Second, they put in significant personal effort to learn new skills related to their interest. Third, the pursuit should have the potential to become a profession or career. Fourth, the participation generates personal and social rewards, like a sense of achievement, which contributes to their wellbeing. Fifth, they often share their enthusiasm with like-minded people within a community of practice, like clubs or associations. Lastly, their involvement can impact the shaping of their personal or social identity (Stebbins, 1982). Of course, people engage at different levels, and not everyone fully meets all the criteria. Some may focus on skill development, while others might value the social connections or social identity. Therefore, the seriousness varies on a spectrum and allows diverse degrees of involvement.

Serious leisure has been a topic of investigation over the past four decades (Gallant, 2017; Gallant et al., 2013). Scholars from many disciplines, including library and information science, have been exploring different aspects of people's engagement with serious leisure. One of the common questions is about what motivates people to spend so much time and energy staying active in their chosen pursuit for the long term. Research shows that serious leisure is usually rewarding, and personal satisfaction is one of the motivations (Cheng & Pegg, 2016; Brooks et al., 2019). Reaching inner peace is also one of the motivations (Heo et al., 2021), as they immerse themselves in the pursuit they like and can experience a state of flow. Flow happens when someone immerses themselves in a delicate and challenging task, losing track of time because of experiencing joy (Csikszentmihalyi, 1990; Heo et al., 2010; Dieser et al., 2015; Standridge et al., 2020; Taylor & Carr, 2023; Larson & Sri, 2024). Scholars in this area explored a broad range of hobbies, such as knitting (Brooks et al., 2019; Mansourian, 2021e), gardening (Cheng & Pegg, 2016), singing (Heo et al., 2021), climbing (Carter, 2023), dancing (Joseph & Southcott, 2019; Finkielsztein, 2023), birdwatching (Mansourian, 2021d; Randler, 2023a), cooking (Marine-Roig et al., 2023), mountain biking (Taylor & Carr, 2023), surfing (Cook, 2022), pottery (Mansourian, 2022b), etc.

In recent years, a few research fronts have emerged in this area. One front explores conceptual connections between leisure and other concepts on a wide spectrum ranging from existential notions, such as meaning and purpose (Iwasaki et al., 2018; Kono et al., 2018; Joseph & Southcott, 2019; Cook, 2022; Iso-Ahola & Baumeister, 2023), to culturally specific perspectives, such as the Japanese concept of *Ikigai* (Kono et al., 2019; Kono & Walker, 2020; Seko & Hirano, 2021; Douziech, 2024). This line of inquiry can deepen our understanding of experiencing leisure through observing its relationship with experiences and culturally embedded views. It seeks to move beyond conventional definitions of leisure, which often focus on relaxation, amusement or recreation. Therefore, leisure scholars are increasingly investigating how leisure contributes to our overall wellbeing and sense of purpose in life. The integration of culturally embedded perspectives, such as Ikigai, which encapsulates a sense of purpose in daily life, offers new views into how different societies see leisure in their cultural contexts.

At the same time, exploring information experience as a personally meaningful activity is a recent area in information behaviour scholarship (Gorichanaz, 2019; Gorichanaz & Latham, 2019). This emerging area shifts the focus from traditional information seeking to a more holistic view that considers how people engage with information in the ways that are embedded in their lived experiences. In my research, I have integrated these approaches to develop my own perspective, which brings together the study of leisure, sense-making, and information experience with a phenomenological perspective. I adopted a phenomenological perspective (Van Manen, 2016) because I intended to explore the lived experiences of bonsai growers as they engage in this hobby. By doing so, I aimed to explore how people interact with information in ways that extend beyond utilitarian purposes, as part of a broader existential and affective experience. Therefore, among the possible approaches to studying information behaviour in this context, I used information experience, which has a phenomenological flavour. Christine Bruce and her colleagues introduced this approach in 2014 (Bruce et al., 2014). After that, other scholars in the information behaviour area developed the concept and used it in their studies in different contexts (e.g., Gorichanaz, 2019; Hoyte, 2019; Solhjoo et al., 2022; Yu & Liu, 2022; Solhjoo et al., 2024).

On the other hand, companion animals play significant roles in many serious leisure pursuits, where they enrich leisure experiences by engaging in activities such as play, performance or competition alongside humans (Dashper, 2018; Young & Carr, 2018; Danby et al., 2019). For example, equestrian sports need years of training and bonding between rider and horse, forming both a physical and emotional partnership (Luxmoore, 2008; Dashper, 2017; Ford, 2019). Similarly, birdwatching requires developing observational skills and ecological knowledge, as birders follow migrant patterns and learn to identify species by sight or sound (Randler, 2023b; Lundquist et al., 2025). Nonhuman animals in these contexts are not passive. They are active participants and partners. They contribute to shaping the experience, and their presence adds emotional depth to these experiences, which enhances the overall value and impact of the leisure pursuits (Carr, 2016; Dashper, 2018; Danby et al., 2019).

The findings and discussion that follow are organised into three sections (the human, the plant, and the in-between) to illustrate how the practice and art of bonsai growing embody specific skills and information experiences. The findings align with the multispecies information science perspective and suggest that to gain a deeper understanding of bonsai growing as a serious leisure practice, we need to recognise nonhuman agency in this context.

The Human

Engaging in bonsai growing, as a hobby or an artistic endeavour, creates a gradual and dynamic process of shaping and being shaped, where a plant–person partnership has a central position in the process. Bonsai growers meticulously prune and wire the branches and roots to shape the trees based on the aesthetic principles of this art. However, dedicating so much time and attention to this delicate practice influences their lifestyle and the way they live. They guide the growth patterns of their bonsai, but the tree also directs their attention and the approaches. That is how a plant–person bond gradually comes into existence. Therefore, while the hobbyist or artist forms the tree, the tree forms the grower too. The slow and deliberate care for the bonsai trees teaches them patience, attentiveness, and a deep awareness of the natural world.

In the next stage, bonsai growers adapt their actions based on the responses they receive from their trees. However, these responses are only visible to their trained eyes and remain invisible to outsiders. Other people who do not have experience in this art do not notice the signs and signals. In this sense, this relationship is mutual, where both are interacting and affecting each other. The grower learns to become more attuned to the tree's rhythms, while the tree influences the grower's mindset. The process reveals that bonsai growing is not an act of control but rather a mutual influence which manifests the bond between the tree and the hobbyist or artist.

From an information science perspective, I found that various forms of information play a crucial role in evolving this bond and rapport. It includes *embedded* and *embodied* types of information (Bates, 2006). Embedded information refers to the information that is inherent in the bonsai tree itself and how it looks. The tree, as a living organism, communicates a wealth of information about its needs. Through careful observation, bonsai growers learn to interpret the signs, such as the appearance of leaves and branches. These forms of information are embedded in the tree and its context. Growers recognise these fine signals and adjust their care accordingly. The tree structure, habits, and positioning of its branches all contain embedded information that guides the grower for the next actions.

Embodied information, on the other hand, involves the knowledge that the grower has within their body. This knowledge is developed through hands-on practice, where their senses, including touch, sight, and even their intuition, become attuned to the needs of the tree. Their embodied knowledge allows them to respond to the subtle signals that cannot always be articulated verbally. Together, embedded and embodied information create a feedback loop between the grower and the tree. The grower brings their accumulated knowledge and skills to the practice, while the tree provides information that directs the grower's response. This dynamic interaction highlights how information, in its various types, is integral to the process of shaping and being shaped in this ongoing relationship and partnership.

While this chapter focuses on the intimate and evolving relationship between bonsai growers and their trees, it is important to acknowledge that it does not happen in isolation. Especially for growers outside of a Japanese cultural context, the process of cultivating a bonsai partnership is shaped by broader information behaviours, including interaction with fellow enthusiasts, mentors and online communities. Bonsai growers engage in various forms of information seeking and sharing, such as informal conversations at club meetings, workshops, or participation in digital forums. These information activities contribute to how growers come to understand and embody the aesthetic and philosophical dimensions of bonsai practice. Thus, the bond between the grower and their trees is shaped by personal care, learning from others and sharing knowledge within the bonsai community.

The Plant

In the art of bonsai, the tree plays an active and autonomous role in the whole process. As a living organism, it has its own needs and desires. Therefore, it responds to environmental elements such as seasonal changes or the grower's interventions. It is not a static material but a living presence that continues to grow and adapt over time. The tree significantly contributes to the final form through its biological reactions and responses. Moreover, each tree species has its own growth habits and limitations. Some trees thicken their trunks slowly, while others respond faster to pruning. The direction and strength of new branches, the emergence of buds, and the ageing of bark depend on their natural processes. These elements cannot be entirely controlled by the grower; instead, they must be understood and respected.

Furthermore, a bonsai's overall health determines how far it can be shaped. Without proper care, such as adequate water, sunlight, and appropriate soil, the tree cannot maintain its strength, which is necessary to support the desired form. It requires time to recover from cuts and wiring, and its reactions often guide the pace and direction of further shaping. In this way, the tree is not shaped alone; it contributes continually to the evolving product.

This active role of trees aligns with emerging scientific perspectives that challenge traditional views of plants as passive organisms. While plants do not have brains or centralised subjectivity, we can argue for plant personhood based on their autonomy, communication, and intelligence (Trewavas, 2014). Trees communicate not only with each other but also with other species. Wohlleben (2016) describes acacia trees releasing pheromones to warn nearby trees of approaching giraffes, prompting them to produce toxins. Through mycorrhizal networks, fungal connections at roots share nutrients and warnings, acting as a connected community (Wohlleben, 2016). This interconnectedness highlights the dynamic agency of trees and underscores why bonsai growing requires not only artistic skill but also responsiveness to the tree's own living processes.

From the information behaviour perspective, the tree also engages in various forms of information interactions. As a living organism, the tree constantly exchanges information with its environment. It senses changes in light, temperature changes and soil conditions, and responds accordingly. These responses, such as the direction of growth or the shedding of foliage, can be considered as expressions of information. In these ways, the tree tells the grower about its needs and desires through observable cues. Recognising and interpreting these signals is essential to maintain the tree's overall wellbeing in the long term.

Furthermore, the tree's preferences and reactions to the environmental changes create a kind of unpredictability in the art of bonsai. Even with the most careful planning, the outcome finally depends on how the tree responds. This is the time that the tree plays its autonomous role. As a result, new shoots may emerge in unexpected places, or branches may not develop as planned by the grower. These natural variations require the grower to be more flexible. In this sense, the tree becomes a partner in the creative process, which impacts the outcome through its own behaviours.

The In-Between

After conceptualising bonsai practice through the lens of information experience with a phenomenological approach, I mapped the human–plant relationship within the Japanese philosophy, which is the traditional foundation of bonsai. I refer to these interpretations as the *in-between*.

Solhjoo et al. (2024) highlight the in-between in information experience as a dynamic, relational agency where information is distributed through engagement among living and non-living entities, such as a cat, a human, and their shared home. In a multispecies context, information experience is not confined to the individual; rather, as the authors note, 'it is not possible to fully experience information without others' (p. 722). The *in-between* reflects mutual dependence and participatory sense-making that arise when humans and non-humans (such as plants) rely on one another and their shared environments to live, adapt, and make sense of the world.

In bonsai growing, this in-between becomes visible through the interaction between the grower and the tree. I found that this engagement manifests some Japanese concepts, including *Kodawary*, *Wabi-sabi*, and *Ikigai*. *Kodawary* means pursuing perfection in a performance or creating a product even though you know perfection is impossible because there is always room for further improvement (Mogi, 2018). However, you accept the impossibility of reaching perfection and enjoy pursuing it. In contrast, *Wabi-sabi* means looking for beauty in imperfection and ephemerality (Pointer Adams, 2017; Kempton, 2018; Wilkinson, 2022). Understanding the coexistence of these paradoxical concepts needs profound comprehension and contemplation. Moreover, implementing these theories into practice requires learning sophisticated skills that can be achieved through several years of apprenticeship under the supervision of a master or mentor.

The idea of Kodawari (Mogi, 2018; Mansourian, 2022a), which is a commitment to continuous improvement, was clearly reflected in the words of bonsai growers. They spoke of their lifelong dedication to the art while acknowledging that bonsai has no final stage or ultimate mastery. No matter how many years you spend on this art, there is always more to learn, especially through living closely with the tree. Over time, bonsai growers can see gradual progress when they develop a relationship with their trees and interpret their behaviours. They try to form the ideal shape of a tree over a long period of time. However, they accept that perfection is unattainable. Nonetheless, they do not interpret this fact as a failure. It even fuels their passion and encourages them to improve their skills further. Their persistent effort exists in tension with the tree's organic autonomy. This is a quiet negotiation where human intention meets natural resistance and where control and surrender coexist. Therefore, the process itself is a source of fulfilment for them, rather than the final product. Their engagement brings them joy, and they often experience a state of flow, which means being fully immersed in the delicate work of pruning and wiring. In these moments, time seems to disappear, and the act of caring becomes meditative or even therapeutic.

The concept of *Wabi-sabi*, another Japanese term, highlights the beauty in imperfection and the natural ageing process. It embraces the aesthetic of the incomplete and the asymmetrical. In bonsai growing, this concept is evident. Bonsai trees are not meant to be perfect in the conventional sense; rather, they embody beauty through their imperfections, such as unevenly spaced branches or weathered bark. The grower's aim is not to create a flawless tree. On the contrary, they try to enhance its natural beauty while honouring its unique characteristics. The bonsai's cracked bark or asymmetrical growth becomes a way the tree informs the human of time's passage, while the human interprets it as artistry. Wabi-sabi helps the grower understand that beauty does not exist in perfection, but it is in the process of becoming. For instance, Hoare (2019) reflects on how the persistence of moss over centuries in Japanese gardens embodies Wabi-sabi, reinforcing patience and evoking a sense of timeless existence. Therefore, the beauty exists in the natural shapes that emerge over time. In this way, Wabi-sabi challenges conventional ideas of beauty.

The paradox between perfection and imperfection is central to the art of bonsai growing. On one hand, the grower seeks to shape the tree in a way that reflects a sense of harmony. On the other hand, the tree itself is constantly evolving, and the grower must accept that this process is not linear. There is a delicate balance between an ideal form and accepting the tree as it is. The tree may never fully conform to the grower's vision, and this is where the paradox manifests. The pursuit of perfection in bonsai is not about achieving a static state but about making a relationship with the tree that allows it to grow and reveal its beauty.

Ikigai is the third Japanese concept that I conceptualise for the human–plant relationship. Ikigai, which is often translated as 'a reason for living,' emerges from the overlap between passion and purpose. In the context of bonsai growing, Ikigai manifests in the grower's interest and passion for taking care of the tree. It is a source of personal fulfilment for the grower. The process of caring creates a sense of purpose. Caring for each tree is like a unique journey, where the grower's passion, their skill, and their understanding of the bonsai philosophy form a lifestyle. Through this practice, the bonsai's survival depends on the grower's care, while the grower's Ikigai depends on the tree's responsiveness over a long time. Hackney's (2024) study with bonsai practitioners supports this idea, showing that intense human–nonhuman collaboration (e.g., shaping trees) deepens ecological sensitivity. Bonsai itself becomes a mentor, fostering animistic perceptions and pro-environmental attitudes.

Conclusion

Bonsai growing is not just about planting and shaping a tree in a miniature size. It is about understanding the plant's nature and responding to its needs and preferences. It also means adapting an approach to suit the characteristics of the tree. This ongoing interaction creates a bond between the bonsai and the bonsai grower, as the grower becomes attuned to the tree's life cycle. Growers often speak of their trees as partners in this process, where success is not measured solely by the appearance of the tree. For them, the act of caring for the tree creates a sense of respecting the autonomy of the tree, and it requires patience and understanding rather than control. As a result, they experience a multifaceted relationship with their trees which goes beyond the simple act of caring for a plant.

Bonsai art is a journey of constant observing, caring and learning. As they work to shape the tree, they also shape themselves. In bonsai, Kodawari reflects the continuous refinement of the tree, where the grower engages in an iterative, embodied dialogue with the plant, adjusting, pruning, and responding to its growth patterns. The grower accepts the fact that there is no absolute excellence and ultimate mastery in this art. There is always room for further improvement. Therefore, engaging in this art is like a lifelong journey rather than producing a product and finishing it. Growers often find that their aesthetic sensibilities evolve as they refine their understanding of the natural world. They begin to appreciate details that may have once seemed insignificant. This careful observation nurtures a connection with nature and enhances their appreciation for the beauty of imperfection. In this way, bonsai growing becomes a form of personal transformation, where the pursuit of perfection is not all about the tree. It is also about the grower's aesthetic maturity as well.

The emotional bond between bonsai growers and their trees is powerful and personal. Many growers describe a sense of attachment. They speak of the tree as a living entity with its own personality, and the emotional connection becomes stronger over time. This bond is not just about admiration for the form. It includes the joy and satisfaction that come from caring for a living organism over a very long term. The emotional dimensions of this connection are attached to a sense of fulfilment. Eventually, the trees become more than a plant for them. It will become a companion that offers a sense of stability and grounding. This connection influences the grower's identity, which can shape their sense of self as someone who is patient and connected to nature. For many, bonsai growing is more than a hobby. It is a lifestyle that defines their identity.

Through the lens of Ikigai, the tree depends on the grower's care, and the grower finds fulfilment in nurturing the tree. It also manifests a sense of mutual reliance and meaning. At the same time, the concept of Wabi-sabi teaches them to accept imperfection and ephemerality. The paradox between seeking perfection and accepting imperfection is not a contradiction. It is a harmony that highlights the practice and shows how the grower and the tree evolve through an ongoing process.

The role of information, both embedded and embodied in the grower's experience and the tree's responsiveness, enriches this relationship. Immersing in this process creates a feedback loop where knowledge is constantly exchanged and refined. The grower's actions are influenced by the signals from the tree. On the other hand, the tree itself responds to the care it receives. This mutual relationship shapes the physical form of the tree and the mindset of the grower. Ultimately, bonsai growing is more than just a horticultural practice. It is a philosophical journey. It teaches us to balance ambition with acceptance. It shows how precision and patience work and how we can control and surrender at the same time.

References

Bates, M. J. (2006). Fundamental forms of information. *Journal of the American Society for Information Science and Technology*, *57*(8), 1033–1045.

Brelje, K. (2023). Caring for plants: Care ethics in relationships between human carers and plant cared-for (Doctoral dissertation, Temple University). ProQuest Dissertations & Theses Global. https://www.proquest.com/docview/2912161925

Brooks, L., Ta, K.-H. N., Townsend, A. F., & Backman, C. L. (2019). "I just love it": Avid knitters describe health and well-being through occupation. *International Journal of Retail & Distribution Management*. https://doi.org/10.1177/0008417419831401

Bruce, C., Davis, K., Hughes, H., Partridge, H., & Stoodley, I. (2014), *Information experience: Approaches to theory and practice*. Emerald Group Publishing.

Carr, N. (2016). *Domestic animals and leisure*. Palgrave Macmillan.

Carter, D. P. (2023). Sustaining civic recreation: Understanding what shapes climbers' contributions to local resource stewardship and advocacy. *Journal of Outdoor Recreation, Education, and Leadership*, *15*(1), 77–93. https://doi.org/10.18666/JOREL-2023-11642

Chan, P. (2018), *The bonsai bible: The definitive guide to choosing and growing bonsai*. Mitchell Beazley.

Cheng, E. (H.-P.), & Pegg, S. (2016). "If I'm not gardening, I'm not at my happiest": Exploring the positive subjective experiences derived from serious leisure gardening by older adults. *World Leisure Journal*, *58*(4), 285–297. https://doi.org/10.1080/16078055.2016.1228219

Cook, P, (2022). Pleasurable surfing is possible: Ethnographic insights into the constructive sociation choices behind meaningful nothingness. *Leisure Studies*, *41*(6), 816–830, https://doi.org/10.1080/02614367.2022.2085773

Csikszentmihalyi, M. (1990). *Flow: The psychology of optimal experience*. HarperCollins.

Danby, P., Dashper, K., & Finkel, R. (2019). Multispecies leisure: Human-animal interactions in leisure landscapes. *Leisure Studies*, *38*(3), 291–302. https://doi.org/10.1080/02614367.2019.1628802

Dashper, K. (2017). *Human-animal relationships in equestrian sport and leisure*. Routledge, Taylor & Francis Group.

Dashper, K. (2018). Moving beyond anthropocentrism in leisure research: Multispecies perspectives. *Annals of Leisure Research*, *22*(2), 133–139.

Dieser, R. B., Christenson, J., & Davies-Gage, D. (2015). Integrating flow theory and the serious leisure perspective into mental health counselling. *Counselling Psychology Quarterly*, *28*(1) 97–111.

Douziech, J. (2024). The Ikigai framework: Supporting meaning in life. *Journal of Policy and Practice in Intellectual Disabilities*, *21*(2). https://doi.org/10.1111/jppi.12505

Finkielsztein, M. (2023). Between sport and leisure: Competitive senior ballroom dancing as serious leisure. *Qualitative Sociology Review*, *19*(4), 52–70. https://doi.org/10.18778/1733-8077.19.4.03

Ford, A. (2019). *Sport horse leisure and the phenomenology of interspecies embodiment*. Leisure Studies, 1–12. https://doi.org/10.1080/02614367.2019.1584231

Gallant, K. (2017). Serious leisure: Past, present, and possibilities. In *The Palgrave handbook of leisure theory* (pp. 437–451). Palgrave Macmillan UK. https://doi.org/10.1057/978-1-137-56479-5_25

Gallant, K., Arai, S., & Smale, B. (2013). Celebrating, challenging and re-envisioning serious leisure. *Leisure/Loisir*, *37*(2), 91–109. https://doi.org/10.1080/14927713.2013.803678

Gorichanaz, T. (2019). Information experience in personally meaningful activities. *Journal of the Association for Information Science and Technology*, *70*(12), 1302–1310.

Gorichanaz, T., & Latham, K. F. (2019). Contemplative aims for information. *Information Research*, *24*(3). http://InformationR.net/ir/24-3/paper836.html

Hackney, D. (2024). Nature-connection for the Anthropocene: A multi-sited exploration of relationships between humans, non-humans, materiality, and discourse (Doctoral dissertation, University of St. Mark & St. John). Marjon Institutional Repository. https://marjon.repository.guildhe.ac.uk/id/eprint/18008/

Heo, J., Lee, S., Im, S., & Bae, J. (2021). "When I sing, I feel peaceful": Experiences of older choral singers from North Korea. *Journal of Applied Gerontology*. https://doi.org/10.1177/0733464820988799

Heo, J., Youngkhill, L., Pedersen, P. M., & McCormick, B. P. (2010). Flow experience in the daily lives of older adults: An analysis of the interaction between flow, individual differences, serious leisure, location, and social context. *Canadian Journal of Aging*, *29*, 411–423.

Hoare, N. (2019). A multi-species ethnography of nature and time: Human's long-standing relationship with moss in the Japanese Temple Garden. *Global Horizons*, *2*(1), 26–43. https://globalhorizonsjournal.wordpress.com/portfolio/a-multi-species-ethnography-of-nature-and-time-humans-long-standing-relationship-with-moss-in-the-japanese-temple-garden/

Hoyte, P. (2019). Implications of the embodied, enactive mind on theorizing about information experience. *Proceedings of the Association for Information Science and Technology*, *56*, 413–416.

Iso-Ahola, S. E., & Baumeister, R. F. (2023). Leisure and meaning in life. *Frontiers in Psychology*, *14*. https://doi.org/10.3389/fpsyg.2023.1074649

Iwasaki, Y., Messina, E. S., & Hopper, T. (2018). The role of leisure in meaning-making and engagement with life. *The Journal of Positive Psychology*, *13*(1), 29–35. https://doi.org/10.1080/17439760.2017.1374443

Joseph, D., & Southcott, J. (2019). Meanings of leisure for older people: An Australian study of line dancing. *Leisure Studies*, *38*(1), 74–87. https://doi.org/10.1080/02614367.2018.1544655

Kempton, B. (2018). *Wabi Sabi: Japanese wisdom for a perfectly imperfect life*. Little, Brown Book Group.

Kono, S., Ito, E., & Gui, J. (2018). Empirical investigation of the relationship between serious leisure and meaning in life among Japanese and Euro-Canadians. *Leisure Studies*. https://doi.org/10.1080/02614367.2018.1555674

Kono, S., & Walker, G. J. (2020). Theorizing Ikigai or life worth living among Japanese university students: A mixed-methods approach. *Journal of Happiness Studies*, *21*(1), 327–355. https://doi.org/10.1007/s10902-019-00086-x

Kono, S., Walker, G. J., Ito, E., & Hagi, Y. (2019). Theorizing leisure's roles in the pursuit of Ikigai (life worthiness): A mixed-methods approach. *Leisure Sciences*, *41*(4), 237–259. https://doi.org/10.1080/01490400.2017.1356255

Larson, R., & Sri, S. (2024). Flow as a catalyst for positive youth development. *Journal of Leisure Research*, *55*(5), 600–620. https://doi.org/10.1080/00222216.2024.2303640

Lundquist, E., Peterson, J., Truong, M.-X., Gumucio, G., & van der Wal, R. (2025). Birdwatching in the digital age: How technologies shape relationships to birds. *Bioscience*. https://doi.org/10.1093/biosci/biaf047

Luxmoore, K. (2008). *Introduction to equestrian sports*. Landlinks.

Mansourian, Y. (2020). How passionate people seek and share various forms of information in their serious leisure. *The Journal of the Australian Library and Information Association*, *69*(1), 17–30. https://doi.org/10.1080/24750158.2019.1686569

Mansourian, Y. (2021a). Bonsai in the time of COVID: The miniature, the social and the solitary. *Cosmopolitan Civil Societies: An Interdisciplinary Journal*, *13*(2), 12–27. https://doi.org/10.5130/ccs.v13.i2.7588

Mansourian, Y. (2021b). Information activities in serious leisure as a catalyst for self-actualisation and social engagement, *Journal of Documentation*, *77*(4), 887–905. https://doi.org/10.1108/JD-08-2020-0134

Mansourian, Y. (2021c). Joyful information activities in serious leisure: Looking for pleasure, passion and purpose. *Aslib Journal of Information Management*, *73*(5), 601–617. https://doi.org/10.1108/AJIM-01-2021-0002

Mansourian, Y. (2021d). "Hide, stay still and listen carefully": Birdwatchers' information seeking and sharing on YouTube. In *Proceedings of the annual conference of CAIS*. https://doi.org/10.29173/cais1191

Mansourian, Y. (2021e). "Knitting keeps my mind sharp and my fingers nimble": Knitters' information seeking and sharing during the COVID lockdown. *Information Science Trends*. https://doi.org/10.5281/zenodo.4885353

Mansourian, Y. (2022a). *Bonsai growing in Australia: Practicing Ikigai and Kodawari within a hobby. In Poster session presented at 4th Symposium of the Japan society for the promotion of science Alumni Association in Australia*, Sydney, Australia.

Mansourian, Y. (2022b). The passion and pleasure of information sharing in pottery practice. In L. Rothfritz, & E. Greifeneder (Eds.), *Proceedings of ISIC: The information behaviour conference Berlin, Germany*, (Special Issue ed., Vol. 27). (Information Research). University of Boras, Sweden. https://doi.org/10.47989/irisic2244

Mansourian, Y. (2024). Information experiences of bonsai growers: A phenomenological study in serious leisure. *Journal of Documentation*, *80*(6), 1645–1659. https://doi.org/10.1108/JD-02-2024-0044

Marine-Roig, E., Lin, M. P., & Llonch-Molina, N. (2023). Cooking classes as destination image contribution: A serious leisure perspective. *Leisure Sciences*, 1–22. https://doi.org/10.1080/01490400.2023.2253806

Mogi, K. (2018). *Awakening Your Ikigai: How the Japanese wake up to joy and purpose every day*. The Experiment.

Norman, K. (2014). *The beginner's guide to bonsai: How to create and maintain beautiful miniature trees and shrubs, shown in more than 230 step-by-step photographs*. Anness Publishing.

Pointer Adams, J. (2017). *Wabi-Sabi welcome: Learning to embrace the imperfect and entertain with thoughtfulness and ease*. Workman Publishing Company.

Prescott, D. (2015). *The Bonsai handbook*. Fox Chapel.

Randler, C. (2023a). Elite recreation specialization and motivations among birdwatchers: The case of Club 300 members. *International Journal of Sociology of Leisure*, 6(6), 209–223. https://doi.org/10.1007/s41978-022-00129-3

Randler, C. (2023b). Motivations for birdwatching: Support for a three-dimensional model. *Human Dimensions of Wildlife*, *28*(1), 84–92. https://doi.org/10.1080/10871209.2021.1993385

Seko, K., & Hirano, M. (2021). Predictors and importance of social aspects in ikigai among older women. *International Journal of Environmental Research and Public Health*, *18*(16), 8718. https://doi.org/10.3390/ijerph18168718

Solhjoo, N. (2026). Research pathways for multispecies information science. In N. Solhjoo (Ed.), *Multispecies Information Science* (pp. 91–110). Routledge.

Solhjoo, N., Krtalić, M., & Goulding, A. (2022). Pets and people: information experience of multispecies families. *Journal of Documentation*, *78*(5), 1092–1108. https://doi.org/10.1108/JD-03-2021-0052

Solhjoo, N., Krtalić, M., & Goulding, A. (2024). The in-between: Information experience within -humancompanion- animal living environments. *Journal of Documentation*, *80*(4), 701–729. https://doi.org/10.1108/JD-08-2023-0160

Standridge, S. H., Dunlap, R., & Hamilton, G. (2020). Retirement and flow: Can the casual leisure pursuits of older adults in retirement create the experience of flow? *Activities, Adaptation, & Aging*, *44*(3), 192–209. https://doi.org/10.1080/01924788.2019.1651177

Stebbins, R. A. (1982). Serious leisure: A conceptual statement. *Pacific Sociological Review*, *25*, 251–272.

Stebbins, R. A. (1992). *Amateurs, professionals and serious leisure*. McGill-Queen's University Press.

Stebbins, R. A. (1997). Casual leisure: A conceptual statement. *Leisure Studies*, *16*(1), 17–25.

Stebbins, R. A. (2001). *New directions in the theory and research of serious leisure*. Lewiston, Edwin Mellen.

Stebbins, R. A. (2007). *Serious leisure: A perspective for our time*. Transaction.

Stebbins, R. A. (2020). *The serious leisure perspective: A synthesis*. Palgrave Macmillan.

Stebbins, R. A. (2021). *Leisure lifestyles: Organizing everyday life for fun and fulfilment*. Emerald Publishing.

Taylor, S., & Carr, A. (2023). 'Living in the moment': Mountain bikers' search for flow. *Annals of Leisure Research*, *26*(2), 285–299, https://doi.org/10.1080/11745398.2021.1974906

Trewavas, A. (2014). *Plant behaviour & intelligence*. Oxford University Press.

Van Manen, M. (2016). *Phenomenology of practice: Meaning-giving methods in phenomenological research and writing*. Routledge Taylor & Francis Group.

Wilkinson, D. (2022). Wabi-sabi: A virtue of imperfection. *Journal of Medical Ethics*, *48*(11), 937–938. https://doi.org/10.1136/medethics-2021-108074

Wohlleben, P. (2016). *The hidden life of trees: What they feel, how they communicate.* Greystone Books.
Young, J., & Carr, N. (2018). *Domestic animals, humans, and leisure: Rights, welfare, and wellbeing*, Routledge.
Yu, L., & Liu, Y. (2022). Information experience as an object of LIS research: a definition based on concept analysis. *Journal of Documentation*, *78*(6), 1487–1508. https://doi.org/10.1108/JD-10-2021-0195
Zhao, Q. (2012). *Penjing, the Chinese art of bonsai: A pictorial exploration of its history, aesthetics, styles and preservation.* Better Link Press.

12

DISTRIBUTED AGENCY AND MATERIALITY IN BIODIVERSITY CITIZEN SCIENCE

Björn Ekström and Ola Pilerot

Introduction

Practice theory strands are discordant when it comes to how to treat agency and materiality. According to Silvia Gherardi (2017, see also 2020), there is by now an established distinction regarding how materiality ought to be considered when reasoning about practices—often contrasted as the human-centred and the posthumanist approach. In the former approach, materiality simply mediates human activity, whereas in the latter, materiality is constitutive of practice (Gherardi, 2017). From the posthumanist view, materiality is not solely an element among many others. Rather, practices are understood as shaped through material qualities, whether analogue or digital. Posthumanist practice theory off-sets the view that social theory necessarily starts and ends with human beings. Where a human-centred approach focuses on human subjects and their interactions with an external world, a posthumanist perspective acknowledges symmetrical relations between humans and nonhumans (cf. Gherardi, 2020; see also Law, 1992). Posthumanism provides an outlook where the capability to act is not delimited to human beings but can be afforded to all species alike.

This conceptual diversion between structure and agency, and the role of the material, is also present in practice-based information studies. Pilerot et al. (2017) identify four clusters of theoretical sub-families commonly referred to in research on information practices. These clusters include learning theories, philosophy and sociology, LIS and, finally, contributions in the field of science and technology studies (STS). Some of these STS theorists are most notably aligned with the notion of materiality as integral to practices and have pioneered posthumanist theory together with feminist and organisation studies

DOI: 10.4324/9781003583424-14

researchers (Haider & Sundin, 2023). It is in the vein of these scholars that the present chapter addresses issues of distributed agency and materiality in research on information practices.

This chapter serves to acknowledge and explore the significance of a posthumanist theoretical perspective to practice-based information studies by adopting a stance where humans, other animals, plants, networks and material objects are equally constitutive of enacted practices. As a case in point, we turn towards biodiversity citizen science as an empirical field to illustrate and reflect upon posthumanist reasonings. Biodiversity citizen science is an approach to environmental research that gathers an assortment of participants, including people with or without scholarly affiliation, to produce data on species abundances. These species include all forms of life as divided into species groups, from vascular plants, mosses, lichens, fungi and algae to invertebrates, birds, bats, fishes, amphibians, reptiles, and other vertebrates, including, for example, foxes, hares and seals. To allow the networked registration of witnessed species, biodiversity citizen science is also constituted by technical information systems for data submission, sometimes called biodiversity observation networks (BONs) (cf. Peterson et al., 2022), as well as web linkages enabling information transfer between such systems and an assortment of devices and objects utilised to annotate and document species witnessed in the field (see e.g., Ekström, 2024).

To underline the significance of posthumanist and multispecies approaches to information practices research, the present chapter draws on a reflective analysis of our previous studies on information practices in biodiversity citizen science. A particular focus is placed on prior work on the Swedish BON Artportalen. Artportalen (en. The species portal) is developed and supported by the Swedish Species Information Centre at the Swedish University of Agricultural Sciences, Uppsala. It is used to amass observed species throughout Sweden and functions as a fundamental source of data and knowledge for county administrative boards, the Swedish Forest Agency and municipalities in their work with nature conservation. Quotations from interview participants have been carefully selected from previous research contributions to illustrate the constitution of biodiversity citizen science information practices with regard to facets of materiality and distributed agency. Our reflective analysis of this empirical material is conducted through a theoretical framework rooted in socio-material practice theory, drawing on posthumanism, multispecies studies and STS. This theoretical approach embraces conceptual notions of entangled activities shaped by humans, more-than-humans and material things to escape anthropocentric ramifications of human-centred perspectives on the study of routinised social activities (cf. Gherardi, 2019; Pilerot & Ekström, 2025).

Throughout the chapter, we discuss how a posthumanist approach to practice-based information studies can serve to untangle restrictions of anthropocentric approaches in social thought. More concretely, we emphasise the need for enquiring about the relationality between nature and society that allows thinking about practices as routinised multispecies doings entangled with materialities. The chapter begins by delving into established posthumanist notions of overcoming the subject-object dichotomy between humans and nonhumans. Next, we discuss who and what has the aptitude to act and how distributed agency can be unfolded. Then, we explicate how interconnected material arrangements invoke practices. This is achieved through a reflection on the routinised information-related activities carried out among geographically dispersed citizen scientists in relation to species abundances, as well as applications and BONs in use as part of biodiversity citizen science in Sweden. In this chapter, we survey modes of enquiry in empirical-laden information studies and point towards further posthumanist-orientated explorations of empirical settings where multiple species dwell.

Multispecies Relations in the Enactment of Practice

When Orlikowski (2007) discusses the socio-material approach to practices, she does so in the sense that 'the social and the material are considered to be inextricably related—there is no social that is not also material, and no material that is not also social' (p. 1437). The question then is how to pivot the alignment of enquiry in information studies from merely concentrating on routinised human activity to one where symmetrical agency is acknowledged. In this subsection, we discuss an assortment of aspects of the relationship between humans and nonhuman species. We start by reflecting upon how this relationship can be traced in the setting of biodiversity citizen science, and subsequently interweave theoretical observations with illustrative empirical examples.

It is significant to note, already at this point, that the human registration of reported species—in the ponds of a nature reserve, in the sky on a cloudy day or in the bushes next to a mall—by default is an anthropocentric conception. Even though the human ambition may be to provide a full-fledged view of flora and fauna by way of distributed monitoring for nature preservation and conservation purposes, as noble as this endeavour is, it takes place in a setting where the species observations wouldn't amount to very much without the species themselves. The skylarks seen from afar during winter migration, the squirrel scattering up a tree on sight, the seal staring back with wonder from ashore; they are not so much facets of a practice as they are dwellers in it.

Agency, as will be discussed later in this chapter, is, through a posthumanist perspective, distributed among the actors involved in the practice (Gherardi, 2019). The skylark's winter migration comprises a practice in the sense of an activity occurring routinely (twice a year) through bodily movements (flying) and general understandings (of where to head). In turn, this migration ties into the information practice of observing through the human, and perhaps a pair of binoculars, as catalysts, forming a nexus of practices (cf. Hui, 2017).

By exploring the role of the nonhuman in practice-based information studies, there is ample room to consider Braidotti's (2019) notion of the posthuman position, as understanding that humans, too, are members of a species and how that helps shape practices. Through this view, should the migratory patterns change, the observation as an information practice would change as well. This echoes what Tsing (2015) means when she writes that 'humans too have always been involved in multispecies worldmaking' (p. 22). There is thus a symmetrical relationship between humans and the more-than-human.

The human-nonhuman symmetry in empirical research can be (re)traced to depictions of the natural sciences in established posthumanist research contributions. A telling example of changing multispecies relations is found in Haraway's (2008) portrayal of anthropologist Barbara Smuts's PhD studies on baboons in Kenya in the late 1970s. At the time, good research practice for objective science, as Haraway (2008) describes, meant that researchers would make themselves invisible to the species studied to ensure a neutral position towards the researched objects. For this sake, Smuts initially sought to interact as little as possible with the baboons, watching them motionless like a fly on the wall. Adhering to conventions of the natural sciences at the time, Smuts attempted to attain a pure, realistic observation of the primates in their natural habitat.

The effort, as articulated by Haraway (2008), to achieve such detached surveillance did, however, not succeed. Smuts found the baboons clearly react and be on their guard in response to her attempts at standing still and observing their doings out of sight. The primates successively 'inquired if the woman was as good a social subject as an ordinary baboon, with whom one could figure out how to carry on relationships, whether hostile, neutral or friendly' (Haraway, 2008, p. 24). Instead of feigning ignorance of the baboons as an impartial observer, Smuts eventually decided to acclimatise herself to baboon etiquette by changing her posture, manners and vocal expressions, and in doing so, she pursued becoming a subject on par with the species under scrutiny (Haraway, 2008). She sought, and achieved, to overcome the subject-object dichotomy of the researcher versus the species, as the baboons saw her as a trustworthy actor in their vicinity, which allowed her to collect data for her studies (Haraway, 2008). When the baboons called, Smuts responded, and when they regarded her presence as a nuisance, she let them be.

Haraway's (2008) description of Smuts's approach as a way to domesticate herself to the baboons is a fascinating reflection on how the researcher is repositioning herself in her empirical setting. It speaks of a situated activity during which the researcher adjusts her doings around, and relations to, the species of interest. Seen through a practice-based lens, this adjustment is reminiscent of Pickering's (1995) *mangle of practice*: The way that humans, tools, theoretical approaches, activities, and conceptualisations undergo constantly altered relations in scholarly conduct. The alteration occurs on-site, in the case of Smuts, when previous routinised and taken-for-granted activities can no longer be upheld in the effort to ensure that science can be conducted, but are changed and adapted to the situation at hand.

A similar human-species alteration can be recognised when reverting to the case of biodiversity citizen science. While adjusting more to the continuation of species' lives than to their activities, the avoidance of killing and sampling during observation and identifying activities of specimens constitutes such an adjusted activity (cf. Ekström, 2023). A participant in a study on the BON Artportalen (En. *The species portal*) states that

> [w]hen it comes to [killing] insects and the like [...], I think that it is a completely obsolete way of working and that it is immoral, especially when it comes to rare insects. Because today one can, with a camera and zoom lenses and other [tools], come so close to an insect that it is not needed.
> *(Participant Daniella, in Ekström, 2023, p. 160)*

The quote is an illustrative example of a changing facet of the information practices of observing and identifying species. As technical solutions become widespread, ubiquitous and affordable, participants' inclinations to collect specimens for expert retrieval dissolve. Subsequently, the specimens are replaced by the capturing of digital representations for more or less trouble-free networked transfers of species observations to BONs. Activities of capturing recorded wildlife change over time, and so, human–animal relations shift with their actors' moral reflection upon their practices.

Value-related facets to the relations emerging between humans and other species are thereby prevalent regarding how affective bonds are formed. As Braidotti (2013) so poignantly lets us know in her critique of nonhuman species as the other, [a]nimals have long spelled out the social grammar of virtues and moral distinctions for the benefit of humans' (p. 69). Braidotti (2013) exemplifies these attributions through traditional depictions of the eagle as an upright sort, the fox as an untrustworthy character and the lamb as the symbol of innocence. While these aspects stem from folk tales of past times, they tell of the relations formed by humans engaging with nonhuman species.

On this subject, the affective ties between humans and nonhumans have been the focus of enquiry in environmental scholarship. Lorimer (2007) uses the concept of *nonhuman charisma* to delineate 'the distinguishing properties of a nonhuman entity or process that determine its perception by humans and its subsequent evaluation' (p. 915). In so doing, Lorimer (2007, 2015) understands this form of charisma in a relational sense and differentiates between *ecological*, *aesthetic* and *corporeal* charisma: *Ecological charisma* refers to detectability and, more concretely, the ability of a human to differentiate a species from another; *aesthetic charisma*, in turn, relates to the extricate properties that an organism possesses and how those properties relate to affective relations with a human; *corporeal charisma*, moreover, has to do with engagement in practice with species over time (Lorimer, 2007). Taken together, Lorimer (2007) suggests, these modes of charisma can be understood as forming conceptions of flagship species that organise order and bring together a variety of groups—including non-governmental organisations, scientists and volunteers—of biodiversity conservation in the UK.

The concept of *nonhuman charisma* has proven fruitful for the assessment of species' charisma in related contexts, too. Lundquist (2018) makes use of *nonhuman charisma* in her enquiry into the community of birdwatchers and describes how birds possess *ecological charisma* through their relatively distinguishing characteristics. The *aesthetic charisma* is furthermore traced to another participant's descriptions in Lundquist's (2018) study of two species: The Black-tailed godwit and the Lesser white-fronted goose. While a participant considers the godwit good-looking, the goose bears the characteristics of a pretty bird. In the depiction of reasonings with participants, Lundquist (2018) draws on Lorimer's (2007, 2015) concept to further illustrate the two birds' *resplendent* and, respectively, *cuddly charisma*.

What is more, Lundquist (2018) illustrates the epiphany imbued in a study participant's anecdote of seeing a goldcrest, which further spurred the participant's interest in birdwatching. Similarly, the *ecological*, *aesthetic* and *corporeal* dimensions of *nonhuman charisma* can be identified in the case of members of a botanical association trawling the island of Öland's landscape in search of wildflowers (Ekström, 2024). The *ecological* and *corporeal charisma*, for instance, could be traceable to participants gently turning a dandelion to make sense of its identifying features. The *aesthetic charisma*, moreover, can be found in the participants' interactions with orchids and how these exemplars of apparent blooming beauty are present at the harsh limestone ground of Öland.

The acknowledgement of human-nonhuman symmetry is a cornerstone in posthumanist theory and enables, as this section has shown, the identification of nuances in how information is engaged with as well as how species and things are prominent in the enactment of practices. As a continuation of such an adopted perspective, research on the affective ties between humans and nonhumans would provide further knowledge on how species are understood and dealt with in the routinised information-related activities enacted through these relations.

This subsection has broached the symmetry of humans and nonhumans imbued in their relations from a posthumanist view. The depiction of a posthumanist approach to practices, participants' relations to species, and the *mangle of practice* and *nonhuman charisma* has been exemplified through previous empirical research contributions. The chapter's next subsection deals with nonhumans' capabilities to act, through the notion of distributed agency.

Distributed Agency

With the assumption of the human–nonhuman symmetry in mind, this part of the chapter addresses the notion of agency and its distribution among humans and nonhuman species, as well as the material qualities of technical objects and arrangements that, altogether, through doings and sayings, shape the enactment of practices. Or, as Haider and Sundin (2023) write on the subject of socio-material theory along with its relationship to posthumanism, 'humans and nonhumans are seen as actors in the shaping of the world and the focus is on understanding the various ways in which they are entwined' (p. 151). Danielle Celermajer (2020), moreover, traces such a stance to individualist notions of agency and structuralist and poststructuralist theories, as well as theories of new institutionalism and those of collective responsibility. Drawing on the posthumanist and socio-material perspective to explore and analyse information practices in particular, this section embarks on how agency, as the capability to act, can be understood as dispersed and distributed among actors rather than as solely originating from the human subject.

The entwinement of activities—whether expressed as doings or as sayings—is central to a posthumanist approach to practices. For this reason, it is significant to outline the distribution of agency, for example, in the sense of how a 'researcher's Google search is constituted by the performativity of computers, networks, software, algorithms, directories, databases, and infrastructure, as these are enacted by the human agencies entailed in their design, construction, and operation' (Orlikowski, 2007, p. 1445). As such, the accountability of actions is formed through relations and through interactions among multiple actors rather than as only originating from single individuals. That is, groups and objects can, through this view, be understood to act, not by themselves but as a whole. This refers to the relational approach to practices, as previously outlined in this chapter, through which 'agency cannot be understood 'simply' as human action, but rather must be understood as always already configured by structural conditions' (Feldman & Orlikowski, 2011, p. 1242). The aptitude to do and say things is thereby not limited to isolated people but instead formed through networks of people and material objects that shape the practices—the constraints as well as the opportunities—for these very doings and sayings.[1]

BONs such as Artportalen are built and maintained by humans, but they also invoke the understanding of species and their whereabouts through the data that is reported to them. This invocation is enabled through the structured data as shown in tables and as visualised on the Geographical Information System (GIS) maps upon queries made. Artportalen shapes participants' activities as much as the participants' activities shape Artportalen. For instance, as each participant reports species throughout the years, these reports are appended to species lists. These lists, moreover, function as incentives for further participation. Or as one participant puts it,

> [T]he citizen scientist part has [for me] to some extent been replaced by this collector's desire, to use Artportalen as a small bag where you put various [species] and hope that [the bag] will become big and large, which is a purely personal driving force
>
> *(Participant Eric, in Ekström, 2022, p. 257)*

Through the affordance of Artportalen, ongoing participation incites further participation and the endeavour of stockpiling species in a dedicated repository, as made openly available. While the participant in question is shown to be a highly skilled and competent nature enthusiast, the quantification of species as derived from the quote is telling of a view of nature as something that can be collected. The structural conditions (cf. Feldman & Orlikowski, 2011) are here evident as configuring the practices enacted; people, material objects and species are entangled in their enactment of practices (cf. Pilerot, 2014).

So, technical devices and nonhuman systems are prevalent as carriers of agency in the enactment of information practices. As observed during a study of a field excursion by a botanical association in south-eastern Sweden, participants tend to use computer-vision-supported smartphone applications such as Google Lens to assist in the identification of wildflowers (Ekström, 2024). The alignment between the human, a found and observed strain of *Sesleria uliginosa*, and the algorithmic eye of Google Lens imbued the practice with distributed agency. In particular, the smartphone application, through its deep learning framework, has the capacity to (at least try to) successfully determine the strain of grass. The automated species recognition tool and the responses it provides lead to shaping the practice in question as the participants gaze at the screen and try to wrangle the strain to get the application to give them an answer. The inclusion of computer vision in botanical fieldwork structures the activities among participants, material objects and nonhuman species. Such 'social regularities are always "in the making"' (Feldman & Orlikowski, 2011, p. 1242; see also Barad, 2003). Applications thereby have the capability to steer the field-based activities, thereby forming the constitutive relationship between application, nature and humans in a performative sense.

The notion of distributed agency provides a multitude of (non)humans to come together and act. It allows a viewpoint to empirical research where negotiations are made visible, where agreements and consensus are centred and where standards are shaped or not followed at all. Situated activities are relationally tied, through a web of connections (Gherardi, 2006), to the formation of large-scale phenomena, and the other way around. This is certainly the case when it comes to information practices in biodiversity citizen science as well. As a multitude of participants engaged in Artportalen are relationally tied to each other, as well as to the species reported, the reporting activities have bearing on reputation being formed. Or, as Participant Adam states, '[s]ome people who report wildly, they will soon have a reputation that "you cannot be really certain that the person concerned has made a correct identification." You do not want that kind of reputation' (in Ekström, 2023, p. 162).

In other words, as the capability to act is distributed among people, through the information systems that are so central to the practices in question, the responsibility to be correct, honest and trustworthy becomes equally dispersed. Objects of trust, such as photographs and additional metadata, can serve to enhance the trustworthiness of reports, as can metadata annotations such as 'det.' (as in determined), which indicates that a certain report has been validated by an external actor (Ekström, 2023). The notion that the capability to act is scattered among participants here allows the enquiry into values forming through amassed sayings and doings and the exploration of large-scale phenomena from a practice-based perspective. Delving further into the dispersed agency of humans and nonhumans from the perspective of posthumanist practice theory would allow further understandings of credibility-in-the-making.

In this part of the chapter, we have discussed the distribution of agency among human participants, nonhuman species and material objects. It has served to examine how nonhuman agency, in tandem with human agency, is constitutive of practices. From such a notion, as exemplified through vignettes and citations from previous research (e.g., Ekström, 2022, 2023), it is possible to grasp how values imbued in, for instance, biodiversity citizen science are formed by way of report systems as well as how humans tend to utilise them when submitting species sightings. The section has also been devoted to discussing the agency of the algorithmic tool during field-based species observation work, as illustrated by previous research on the observation practices of a botanical association (e.g., Ekström, 2024). The next section gives further attention to the issue of material arrangements and how this notion of materiality can be used as an analytical device in practice-based information studies.

Material Arrangements as Constitutive of Practice

In our discussion so far, we have touched upon aspects of human-nonhuman symmetry and distributed agency. In what follows, we will further ponder facets concerning material arrangements and how these are constitutive of information practices, along with the humans and nonhumans that shape these practices. It is at this point worth returning to the notion of a relational understanding of materiality as referred to at the beginning of the chapter. As Gherardi (2017) states, matter and meaning are 'constituted as relational effects performed in a texture of situated practices' (p. 40). This gives rise to the notion that there is no inherent boundary between the social and the material. Furthermore, it implies what Monteiro and Nicolini (2015) refer to as the silent but essential role of material elements in the enactment of practices. In the following subsection, the material arrangements forming information practices in biodiversity citizen science will be further investigated in relation to field-based practices of observing, annotating and identifying species. We also draw on current research on the role of BONs in the transfer of species reports' particularities to investigate the role of materiality as these information systems shape and reshape such reports.

An illustrative example of material arrangements' bearings for enacted practices can be found by revisiting the previously mentioned empirical case of the field excursion carried out by the botanical association. During the excursion, the associations' participants—in their quest for orchids, plantains, dandelions, and other flowers in bloom—casually strolled along the limestone ground and negotiated the identity of their findings (Ekström, 2024). Common peripherals such as notebooks, smartphones with computer-vision-supported applications, and system cameras were present to document, record, and identify the species witnessed. This accumulation of devices and their collected utilisation by a human invokes another classic technoscience metaphor, which is that of the cyborg.

Especially notable in the context of material objects invoking practices was the excursion leader's assemblage of gadgets. These devices were used to record the path and the observations while simultaneously ensuring that the knowledge, as derived through the geographical location, reached the group (Ekström, 2024). Dictaphone in hand, with a GPS tracker clipped to the belt and equipped with a vest where loudspeakers were fastened to the fabric linings, the leader brought to mind the image of Haraway's (1991) cyborg: A being enhanced by synthetic components. This amalgamation of technical devices, along with the other participants and the barren scenery, altogether gave rise to a variety of information practices which were enacted in relation to the site of enquiry's nonhuman inhabitants: The species.

The leader's device assemblage fostered several enactments of information interactions in the setting. The loudspeakers and a connected headset 'provided a way to sonically follow the events of the tour even if a participant lingered and was not at all times physically located in direct proximity to the leader […]' (Ekström, 2024, p. 11). Likewise, the reproduction and amplification of sound enabled the leader to dwell on details concerning how to spot and differentiate, for instance, one orchid from another. This auditory dispersion of knowledge in situ comprised a performance of information sharing, resulting in further deliberations and negotiations among the participants concerning the species sighted. As the discussions about the landscape prospered, a narrative concerning the space took form, which, in turn, formed the social interaction among the participants. Through ongoing tool use taking place by means of participants tinkering with smartphone applications, cameras and loupes, undecided knowledge objects in a continuous process of definition (cf. Knorr Cetina, 2001) were formed, which, as consensus was reached, were stabilised and subsequently reported to Artportalen.

As illustrated thus far, material objects and their qualities are highly prominent in structuring the enacted routinised activities among practitioners.

From an information studies approach, it is analogous to the exploration of trajectories of information sharing in the scope of a network of design researchers. In his study, Pilerot (2014) shows that material objects as well as information and communication technologies are performative of entangling communities and aligning activities. Moreover, it is highlighted that '[a]nnotated articles, e-mails, the [file-hosting service] dropbox, mindmaps, and post-seminar documentation appearing on the network's website are evidence of information-sharing activities that contribute to constituting practice and to connect' events across time and space (Pilerot, 2014, p. 2013). Material objects are thus—also in the case of biodiversity citizen science—prevalent in the arrangement of routinised information activities since they are co-constitutive of them. Such a theoretical edifice provides a sophisticated approach to follow not only researchers and technological advancements (cf. Latour, 1987), but also to extend this exploration to settings incorporating nonhuman species by considering the entanglement of nature and society.

The intertwinement of material arrangements and practitioners' activities through which practices happen necessitates a deliberation of how the trajectory of information can be explored when reporting species to BONs, but also what happens to the reports when they travel further to other BONs by means of data transfer and aggregation. In the case of biodiversity citizen science, it is relevant to yet again ponder the BONs and how reported species flow between these. As species are reported to these networks, they gain structure through fine-granular and streamlined data entries where participants' annotations are entered into standardised variable columns (Ekström, 2024). The citizen scientists' fieldwork becomes quantified through structuring qualities of information systems and thereby tabulated, where facets of the sightings are fitted into pre-categorised boxes of metadata variables.

Essentially, a BON is a container comprising assorted representations of species as observed in nature. However, when reports are subsequently transferred to other BONs as a means of data aggregation—a process which Peterson and co-authors (2022) call (inter)national connections—the report becomes remodelled. Peterson et al. (2022) describe the journey that the report of a sighted Northern lapwing takes between the sight of observation and Artportalen, and onwards to the biodiversity occurrence system Global Biodiversity Information Facility (GBIF). Besides the lapwing sighting being translated to Artportalen and restructured into variable categories, it undergoes further data moulding when it is subsequently transferred to the GBIF so as to meet the requirements of GBIF's data standards (Peterson et al., 2022). By paying attention to the events during these transfers, it is possible to consider the transmutation of the initial report and how it transforms by way of restructuring through each connection between nodes. Or, as Peterson et al. (2022) write, the sighted lapwing is destabilised through the data moulding occurring during each line of connection, which in turn ruptures the sighting as objective data. Such an approach permits the enquiry into how BONs transform species through connections and how these connections alter the species data amassed.

As much as BONs need to adjust the structure of the reports during their information-related travels across various metadata scaffoldings, negotiations occur in the field as participants try to make sense of what on Earth, so to speak, is living there. As discussed above, these negotiations are conducted through the utilisation of physical objects, but they are also informed by intangible material qualities of the information systems through which the identified observations are reported. Still, not all aspects of the field-based practices are transferable to the BONs. As discussed in the study of the botanical association's field excursion, 'the data produced are regulated but also appended through the data tabulation occurring when storing the reports in the information system' (Ekström, 2024, p. 12). While structure and granularity are gained through tabulation, other doings enacted in the field during identification are lost upon reporting (Ekström, 2024; see also Kasperowski & Hagen, 2022).

However, it is not only the ideal of quantification and calculation that characterises the BONs. It is also possible to trace a gender-related aspect. A recent research contribution on long-term trends of participants' gender and age in Artportalen shows that the BON in question arose from ornithology, which itself is dominated by males, and that the birdwatching culture is replicated in the digital realm (Jönsson et al., 2024). As previously briefly touched upon, the graphical user interface and infrastructural affordances, including, for instance, the accumulated lists of species, 'reproduce values concerned around competing, observing and reporting in large quantities and on the rare and exceptional' (Jönsson et al., 2024, p. 131). This, the authors argue through Lundquist (2018), is likely to be an inheritance from 'a male-dominated amateur culture' (Jönsson et al., 2024, p. 132).

The ambition to report vast quantities or extraordinary sightings is prevalent in several species groups when it comes to Artportalen, as participants seek to stockpile, collect, or curate lists of species as a means of collecting, competing or curating repositories (Ekström, 2022, 2023). The material qualities of Artportalen and the practitioners' incentives to compete, both with each other and themselves, thus work as a parallel appendage that forms the practice. To increase diversity in participation and learning, Jönsson et al. (2024) suggest that managers of citizen science projects ought to involve young people and women to a greater extent. This realisation offers a significant opportunity for information studies research orientated towards posthumanist practice theory to further investigate the information practices and knowledge practices and these, and other, demographic groups (cf. Ekström, 2025), and their doings through the material arrangements shaping their participation.

Concluding Discussion

In this chapter, we have sought to emphasise the relationality between nature and society to guide the reader towards what a posthumanist theoretical perspective can offer for the enquiry into information practices. This has been achieved by the illustrative example of information practices in biodiversity citizen science, as coupled with other contributions on information practices in scholarly conduct and empirical vignettes from posthumanist-orientated research. Overall, the chapter has emphasised the analytical possibilities provided by a theoretical approach emphasising human-nonhuman symmetry, distributed agency, and material arrangements as enacting practice.

As the chapter has shown, a posthumanist approach to practice theory allows overcoming dichotomies of humans and nonhumans, as is emphasised in human-centred approaches to practice, by instead focusing on the symmetrical relationship of such actors. This, in turn, permits a broader understanding of practices by establishing the significance of nonhumans in the enactment of practices and leads to relationality as a point of enquiry. For instance, such a relationality can be traced through enquiries into the nonhuman charisma enacted between multispecies and thereby delve into the formation of affective logics (Lorimer, 2007, 2015; see also Lundquist, 2018). It provides an analytical point of entry to information practice research that would otherwise be difficult to attain.

Moreover, the theoretical notion of distributed agency permits an approach to human-nonhuman co-constitution of practice rather than originating from a single anthropocentric perspective. Yet again, the relational understanding of practices enables the exploration of value-becomings, as discussed, for instance, in relation to dispersed credibility formations in biodiversity citizen science (cf. Ekström, 2023). The emphasis on the structural conditions of practice (cf. Feldman & Orlikowski, 2011) allows exploring the entanglement of agency among various actors. This, moreover, leads to decentring the human subject as an agential carrier and provides insights into the (non)human co-constitution of practice.

At the beginning of the chapter, the human-centred approach to practices was pitted against the posthumanist approach. As has since been discussed, the posthumanist approach acknowledges a theoretical stance of materiality as constitutive of practice, rather than seeking to understand the material only as things that facilitate human activity. The posthumanist approach provides an understanding of how material arrangements and qualities of these arrangements invoke practices. In this chapter, this has been illustrated through the ways that technical information systems such as BONs append, transform and mould species reports, but also the information practices that provide such reports (cf. Ekström, 2024; see also Peterson et al., 2022).

An enquiry into the material arrangements, moreover, enables investigations of how situated activities and large-scale phenomena are intertwined, which methodologically—from a practice point of view—can be explicated through the zooming in and out of practice (Nicolini, 2009). On a similar note, the focus on material arrangements can make relations of inquiry visible and allow the investigation of inherited general understandings of how to act in a given setting, including BONs, and reproduce structural relations in traditionally male-dominated cultures (Jönsson et al., 2024). As discussed throughout the chapter, studying information practices from a multispecies perspective makes it possible to investigate how species (humans, animals, plants, etc.) and material objects co-constitutively shape routinised activities that involve engaging with information.

Note

1 For a critique of this approach to agency in practice theory, see, for example, Schatzki (2025).

References

Barad, K. (2003). Posthumanist performativity: Toward an understanding of how matter comes to matter. *Signs*, *28*(3), 801–831. https://doi.org/10.1086/345321

Braidotti, R. (2013). *The posthuman*. Polity.

Braidotti, R. (2019). *Posthuman knowledge*. Polity.

Celermajer, D. (2020). Distributed agency, responsibility and preventing grave wrongs. *Contemporary Political Theory*, *19*(2), 188–210. https://doi.org/10.1057/s41296-019-00359-3

Ekström, B. (2022). A niche of their own: Variations of information practices in biodiversity citizen science. *Journal of Documentation*, *78*(7), 248–265. https://doi.org/10.1108/JD-07-2021-0146

Ekström, B. (2023). Thousands of examining eyes: Credibility, authority and validity in biodiversity citizen science data production. *Aslib Journal of Information Management*, *75*(1), 149–170. https://doi.org/10.1108/AJIM-10-2021-0292

Ekström, B. (2024). No rose on this one? Citizen science field excursion negotiations at the Great Alvar. *Nordic Journal of Science and Technology Studies*, *12*(1), 5–16. https://doi.org/10.5324/njsts.v12i1.5860

Ekström, B. (2025). Can the exposition of paradata lead to participant diversity in biodiversity citizen science? *Information Matters*, *5*(1). https://informationmatters.org/2025/01/can-the-exposition-of-paradata-lead-to-participant-diversity-in-biodiversity-citizen-science/

Feldman, M. S., & Orlikowski, W. J. (2011). Theorizing practice and practicing theory. *Organization Science*, *22*(5), 1240–1253. https://doi.org/10.1287/orsc.1100.0612

Gherardi, S. (2006). *Organizational knowledge: The texture of workplace learning*. Blackwell.

Gherardi, S. (2017). Sociomateriality in posthuman practice theory. In A. Hui, T. R. Schatzki, & E. Shove (Eds.), *The nexus of practices: Connections, constellations and practitioners* (pp. 38–51). Routledge.

Gherardi, S. (2019). *How to conduct a practice-based study: Problems and methods*. Edward Elgar Publishing.

Gherardi, S. (2020). A posthumanist epistemology of practice. In C. Neesham (Ed.), *Handbook of philosophy of management* (pp. 1–22). Springer International Publishing. https://doi.org/10.1007/978-3-319-48352-8_53-1

Haider, J., & Sundin, O. (2023). Sociomateriality. In A. Hicks, A. Lloyd, & O. Pilerot (Eds.), *Information literacy through theory* (pp. 149–164). Facet Publishing.

Haraway, D. (1991). *Simians, cyborgs, and women: The reinvention of nature*. Routledge.

Haraway, D. (2008). *When species meet*. University of Minnesota Press.

Hui, A. (2017). Variation and the intersection of practices. In A. Hui, T. R. Schatzki, & E. Shove (Eds.), *The nexus of practices: Connections, constellations and practitioners* (pp. 52–67). Routledge.

Jönsson, M., Kasperowski, D., Coulson, S. J., Nilsson, J., Bína, P., Kullenberg, C., Hagen, N., van der Wal, R., & Peterson, J. (2024). Inequality persists in a large citizen science programme despite increased participation through ICT innovations. *Ambio*, *53*(1), 126–137. https://doi.org/10.1007/s13280-023-01917-1

Kasperowski, D., & Hagen, N. (2022). Making particularity travel: Trust and citizen science data in Swedish environmental governance. *Social Studies of Science*, *52*(3), 447–462. https://doi.org/10.1177/03063127221085241

Knorr Cetina, K. (2001). Objectual practice. In T. R. Schatzki, E. Von Savigny, & K. Knorr Cetina (Eds.), *The practice turn in contemporary theory* (pp. 184–197). Routledge.

Latour, B. (1987). *Science in action: How to follow scientists and engineers through society*. Harvard University Press.

Law, J. (1992). Notes on the theory of the actor-network: Ordering, strategy, and heterogeneity. *Systems Practice*, *5*, 379–393. https://doi.org/10.1007/BF01059830

Lorimer, J. (2007). Nonhuman charisma. *Environment and Planning D: Society and Space*, *25*(5), 911–932. https://doi.org/10.1068/d71j

Lorimer, J. (2015). *Wildlife in the anthropocene: Conservation after nature*. University of Minnesota Press.

Lundquist, E. (2018). Flyktiga möten: Fågelskådning, epistemisk gemenskap och icke-mänsklig karisma. [Doctoral dissertation, Stockholm University].

Monteiro, P., & Nicolini, D. (2015). Recovering materiality in institutional work: Prizes as an assemblage of human and material entities. *Journal of Management Inquiry*, *24*(1), 61–81. https://doi.org/10.1177/1056492614546221

Nicolini, D. (2009). Zooming in and out: Studying practices by switching theoretical lenses and trailing connections. *Organization Studies*, *30*(12), 1391–1418. https://doi.org/10.1177/0170840609349875

Orlikowski, W. J. (2007). Sociomaterial practices: Exploring technology at work. *Organization Studies*, *28*(9), 1435–1448. https://doi.org/10.1177/0170840607081138

Peterson, J., Kasperowski, D., & van der Wal, R. (2022). (Inter)national connections: Linking Nordic animals to biodiversity observation networks. In C. Travis, D. P. Dixon, L. Bergmann, R. Legg, & A. Crampsie (Eds.), *Routledge handbook of the digital environmental humanities* (pp. 288–302). Routledge. https://doi.org/10.4324/9781003082798-25

Pickering, A. (1995). *The mangle of practice: Time, agency and science*. University of Chicago Press.

Pilerot, O. (2014). Making design researchers' information sharing visible through material objects. *Journal of the Association for Information Science and Technology*, *65*(10), 2006–2016. https://doi.org/10.1002/asi.23108

Pilerot, O., & Ekström, B. (2025). Information practice and information behaviour: On the need for conceptual and theoretical clarity. *Information Research an International Electronic Journal*, *30*(CoLIS), 461–473. https://doi.org/10.47989/ir30CoLIS52270

Pilerot, O., Hammarfelt, B., & Moring, C. (2017). The many faces of practice theory in library and information studies. *Information Research*, *22*(1). http://www.informationr.net/ir/22-1/colis/colis1602.html

Schatzki, T. R. (2025). Agency. *Information and Organization*, *35*(1), 100553. https://doi.org/10.1016/j.infoandorg.2024.100553

Tsing, A. L. (2015). *The mushroom at the end of the world: On the possibility of life in capitalist ruins*. Princeton University Press.

13

HABITATS OF ARCHAEOLOGICAL KNOWLEDGE

From Information Ecologies to Information-in-Ecologies

Isto Huvila

Introduction

The human past is far more than merely a human one. Throughout the millennia, humans and the other species inhabiting the Earth have been intricately interwoven—both enmeshed (Morton 2010) and entangled (Tsing et al. 2017; Hodder 2012)—into each other in multiple ways (Oma and Goldhahn 2020) that are at the same time self-evident and difficult to grasp. This is perhaps especially difficult in the context of the conventional information science research that continues to be empathetically centred on human beings and their undertakings to an extent that describing human engagements with information as *human* information behaviour has been found to be unnecessary (Bates 2017; Greifeneder and Schlebbe 2022). Even if the engagements with theorising that refuse to make *a priori* divisions between the human and nonhuman have opened for alternative understandings of the configurations of informational ecologies, changing perspective is not easy. It is especially difficult in the contemporary information-rich contexts where alternatives are easy to ignore. Giving nonhuman framings of informational undertakings a greater chance is easier when approached through looking at something we can only know through multispecies and more-than-human enmeshments. One such context is the past before written evidence.

In this chapter, I take a few steps in this direction, drawing on my previous research on archaeological information work. The chapter engages with the archaeological *ecology of information practices* (drawing on Stengers 2005; Huvila 2009) and the concept of *making and taking information* (Huvila 2022) that provide a useful alternative to framing the conventional flow of information from producers to users, whether humans or nonhumans. This approach

DOI: 10.4324/9781003583424-15

acknowledges that neither contemporary nor past human habitats—the latter framed and investigated by archaeologists as archaeological sites—are not only human spaces but also territories and ecosystems inhabited by manifold living and non-living agencies and species involved in making and taking multi-modal information.

From the outset of the contemporary practices of making and taking of archaeological information, the aim of the chapter is to provide an insight into archaeology as an example of an informational enterprise where the multispecies—and, as is evident in this chapter, in broader terms, more-than-human—nature of information practices is particularly apparent. In parallel, the chapter draws on archaeological practice to provide stratagems for approaching other, both obviously and less apparently, multispecies and more-than-human information habitats that information scientists are studying. The discussion draws on an exploration of the habitats of archaeological knowledge. It traverses the entanglements of human and nonhuman agencies in the archaeological record—understood as the material, non-written evidence of the past investigated by archaeologists (Patrik 1985; Lucas 2012)—and knowledge of the human past, and how an inquiry into nonhuman entities contributes to understanding past knowledge-making activities. This chapter enquires specifically into what can be termed archaeology-proper rather than metaphorical archaeologies (Woźny 2021), the latter typically contending with figurative digging and emphasising the materiality and layered nature of evidence. Especially when it comes to its multispecies and more-than-human nature, there is much more to engage with in archaeology beyond such archetypal *topoi*. Such an approach can help to open the black box of the multispecies more-than-human information world beyond what otherwise might risk appearing only half-open, or, as Winner might regret, 'a remarkably hollow one' (Winner 1993, p. 375).

Making and Taking Information: A Multispecies Perspective

The notions of making and taking draw from an observation of how processes that are described as linear flows from information creators to users are often, in practice, plagued by disconnects, movement in multiple directions, and inclusion of participating entities far beyond those assumed and explicitly articulated (Huvila 2022). The conceptual apparatus, or theory, of making and taking engages with one of the key transitions, or trading zones, in the 'red thread of information' (Bates 1999, p. 1048), the one between information provision and use. It stems from observations in a series of studies of archaeological information that direct attention to that information that is often created and provided for imagined audiences, which do not necessarily correspond very closely with actual publics. Correspondingly, instead of merely using pre-packaged, ready-to-use information, information users often need to put considerable effort into getting the information to work for them. The notion of *information making* refers to the diverse major and mundane practices of crafting, creating, and generating information, and *information taking* to corresponding to the, similarly manifold practices of *situational appropriation* (Huvila 2015, 2019d), of how diverse actors take information to use as their own by crafting it by different means to be useful for their specific purposes in a given situation.

The advantages I find in terming information generation as 'making' include its emphasis on the prosaic, craft-like nature of making information happen rather than it being a result of a production process or creative undertaking, and how it places emphasis on mak*ing* rather than individual makers. Taking has comparable affordances that encompass information discovery, acquisition and situational appropriation of already existing information with a focus on how information is crafted for oneself before it is used (as per information use), that is, put into action to contribute to understanding, learning or, for example, sense-making. Instead of underlining makers or takers and making or taking *of* information, the approach follows the rationales of relational ontology (cf. Haraway 2003; Barad 2007) by focusing on how making and taking happen *with* human and nonhuman others. The 'with' itself is not a static relation, but as Cahoon (2025) suggests, by the term 'withing,' it is itself a dynamic practice in the making.

Rather than merely attempting to point out that the continuum of information is complex and in a constant flux, theorising information work in terms of making and taking approaches is something akin to what Whitehead suggested of experiences of being made up of complexes of 'actual occasions' (Whitehead 1978, p. 15). Making and taking are not complex in general, but in a very distinct manner. Information becomes information in a particular sense only at the being or becoming of each specific occasion.

Focusing on making and deemphasising makers helps making and taking also to avoid some of the pitfalls of premising human-centricity in information practices. It requires, however, a parallel shift from framing information ecologies and ecosystems as (as per humans) socio-technical assemblages (e.g., Introne et al. 2024) to conceptualising them as more-than-human agencements that are deeply *enmeshed*, as for Morton (2010), or *entangled*, as for archaeologist Hodder (2012) and anthropologist Tsing et al. (2017). Ecologies entail not merely people together with technologies, information and practices but comprise the complete more-than-human sphere of existence where the actors can *a priori* be of any human or nonhuman kind (cf. Huvila, Enwald et al. 2022). This form of symmetricity does not mean that all agency in the multispecies more-than-human world would be the same, but that all things should initially be approached as potential actors and beings equal in having agency (cf. Latour 1988; Amsterdamska 1990; Pickering 1995) in relation to tools, information and a particular situation at hand.

In a comparable manner, in the continuum of making and taking that iterates indefinitely from one actor to another (or one actor to the same actor) and

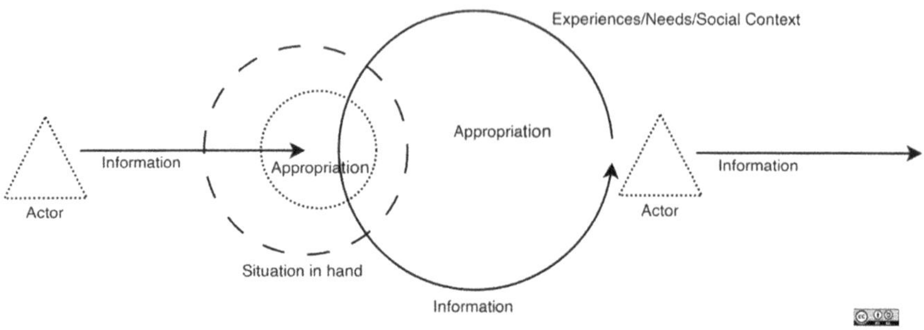

FIGURE 13.1 Making and taking information in a multispecies, more-than-human sense. Figure licence: CC-BY-SA-4.0.

from one instance of making and taking to another, the possible actors are not only humans or machines but can belong to any species (Figure 13.1 cf. (Huvila 2022), which is an iterative rather than a linear process from actor to actor). Work and information work are not only concerns of human actors but also extend to others to an extent. Different more-than-human agencies' diverse engagements lean themselves to be productively conceptualised in terms of work (cf. (Huvila 2018); Figure 13.1. in (Huvila 2009). The situation in hand (in Figure 13.1) is neither a nonhuman context for human engagements nor a more-than-human entanglement in time within which information is made and taken. As argued elsewhere (Huvila 2019c), rather than being a container for activity, I am inclined to see it as a centred set of diverse facets playing out in a specific moment and space in time.

Archaeological Record as a Multispecies Record

Considering the affinities of making and taking to foreground situatedness, illinearity, socio-materiality and the more-than-human nature of information practices and their roots discussed in the extensive empirical research in archaeological information work, it is hardly surprising that archaeology itself unfolds as an illustrative context for exploring these specific aspects of informational undertakings. Even if typical definitions of archaeology position the discipline as a study of the human past through its material remains (Olsen 2012), the archaeological record is far from being merely a record of the human past. In contrast, even if sometimes forgotten, many central insights in the history of archaeology—including some of the key aspects of prehistoric periodisation and basic understanding of livelihood in the past—stem from careful analysis of human interactions with nonhuman entities (Oma and Goldhahn 2020).

Pitblado et al. (2025) criticise the terms 'object' and 'artefact' conventionally used for human-made portable items in archaeology, for being incompatible with Indigenous ideas of their nature, but also of going against how material stuff is conceptualised in many colloquial Western practices with historic and prehistoric things. This applies also to archaeology itself, which, in spite of its conventional focus on humans, has already for decades acknowledged the value of materials well beyond what reasonably can be termed in the literal sense as an artefact (Lucas 2012). Pitblado et al.'s (2025) term of choice *belongings* solves certain problems associated with the notion, although, as acknowledged by the authors, it raises new issues relating to the emphasis of belonging to someone or something, either items to individuals or communities, or vice versa. Establishing such links can be complicated in human-material relations and risks becoming even more intricate in a more-than-human context. What the terminological discussion and attempts to make such linkages do, however, is to direct attention to the very complexity of the issue of the nature of the things the archaeological record consists of.

Animals and Plants in the Archaeological Record

Even if human-made or human-influenced artefacts (or belongings, as for Pitblado et al. (2025)), features and structures are key categories of evidence for generating archaeological knowledge, the archaeological record consists of a broad variety of evidence of the entanglements of human elements with the more-than-human facets in—using the term of Pritchard (2011)—envirotechnical systems of the past. Some of the evidence is indirect. Fishing hooks are reasonable indications of the presence of fish, even if they theoretically could be used for other things than fishing. The same applies, for example, to things like harnesses, farming equipment, animal sheds, and depictions of animals.

Apart from indirect evidence, the archaeological record contains a lot of remains of nonhuman living species. Animal bones and teeth are typical finds that are retrieved for their capability to shed light on the human past. In favourable conditions, the archaeological record can contain traces of soft tissue and, for example, excrement. Apart from mammals, fish, shellfish, birds, and insects can also be found in archaeological contexts. Many of the key applications of natural science methods in archaeology focus on investigating nonhuman remains (Killick 2015). The history of the established branches of archaeology focusing on animal remains (zooarchaeology or archaeozoology) and plants (archaeobotany) goes back to the mid-1900s (Birch 2018). Such evidence is studied to know more about, for example, what humans ate, how they kept themselves warm, what kinds of domestic animals they kept, how they acquired resources for crafts and industry, and how they travelled and transported goods. At the same time, on the basis of animal remains, it is possible to deduce what types of animals existed in different times, how the past species and sub-species differ from present ones, how they lived and died, and how animals lived together with humans as a part of the past ecosystems.

Archaeological research also recovers evidence of parasites and evidence of animals living not only with but also alongside human beings. Sometimes multiple types of evidence coincide, complementing each other (Wallis 2023). More recently, especially isotope and DNA analyses have gained prominence, pushing forward the understanding of, for example, migration patterns and relationships between past populations of living beings (e.g., Wilhelmson and Price 2017).

There are also a lot of remains of past vegetation and fungi in the archaeological record. Seeds and other types of plant remains, but also organic non-animal artefacts such as textiles and objects made of wood and plants, tell stories of past multispecies engagements. Moreover, the nonhuman does not end with animals and plants in the archaeological record. Archaeological collections contain untold quantities of material objects, including pebbles and fossils, that archaeologists have collected for diverse, often undocumented reasons (Johanson 2018). Sometimes, but not always, such objects have traces of use, they have a specific texture, they fit in a human hand, or their presence or position at the site cannot be explained otherwise than by human interventions. Making the distinctions requires what Goodwin describes as professional vision—'consist[ing] of socially organized ways of seeing and understanding events that are answerable to the distinctive interests of a particular social group' (Goodwin 1994, p. 606). For example, Olsson and Lloyd (2017) describe how a geologist's observation can turn natural stones into archaeologically interesting material by showing how they are not local to an investigated site but have been transported there from afar. As Johanson (2018) points out, while the (human) use of many natural objects is impossible to ascertain, refusing to acknowledge the possibility means that perceiving such use is *a priori* rejected.

Nonhuman Agency on Past (More-Than-) Human Beings

If distinctions between archaeological and non-archaeological things are difficult to make when the focus is on human activities, they turn even more delicate when the scope is broadened to the multispecies and more-than-human past. The traces of human exploitation of the nonhuman sphere in the archaeological record are only a part of how it evinces the more-than-human past. As Hamilakis and Overton (2013) emphasise, the distinction between exploitation and interactions is meaningful. Even when the focus of attention is on humans, their only possible relation to the more-than-human is not that of exploitation. The human associations with the more-than-human sphere are far more complex. Even if animals are hunted, hunting is not only about consumption. The same applies to the domestication of animals. Archaeology helps to understand not only what humans ate and what materials were used to make artefacts, but also what humans did and could not eat, how they appreciated the presence and companionship of particular species and things, and how the availability of some materials provided opportunities and limited their possibilities to develop technologies, find and grow nourishment and exploit nature. No artefact or belonging is solely a human creation in material or even in a conceptual sense. Characteristics of animals and plants and the non-living environment, as Gibson (1977) famously underlined, both afford and constrain specific uses.

Sometimes multiple species live in symbiosis. Lucas (2018) uses turf buildings as an example of assemblages of living and non-living elements coming together to form a functioning living whole. While humans are active agents in constructing such buildings, they are certainly not alone in making them happen and sustain. In other cases, living species can also quite actively resist human influence and exert their own agency, both temporarily by going against human intentions. Co-presence of parasitic organisms, diverse otherwise unwanted species and evidence of short-lived attempts to establish human presence in hostile environments evince the occasional forcefulness of the more-than-human agency. Even human remains like bones, teeth, bodies and their fragments and evidence of disappeared organic material like cavities and colourings in the soil indicate the same. Diet, sickness, health, and causes of death all leave marks of engagements in human remains that are not exclusively human but as much remains of the lives of other species. Such traces include remains of parasites (Pye 2021) but also evidence of the impact of the material world on human beings, for example, marks of a falling roof on human skeletons (Wilhelmson and Dell'Unto 2015) or teeth worn by sand that crumbled in bread flour from porous millstones (Leek 1972).

There is also a lot of evidence of natural phenomena in the archaeological record, from volcanic eruptions and earthquakes, but also weather conditions and climate in general. For example, in the Roman city of Pompeii that was struck by an earthquake in 62 CE and covered by ash from a massive eruption of Mount Vesuvius 17 years later, the volcanic nonhuman agency is plain to see. It was obvious already for the antiquarians working at the turn of the 19th century, long before archaeology became a scholarly discipline in our contemporary sense (Heringman 2018). Climate and weather also leave traces in pollen and tree rings in wooden objects, building remains and natural wood preserved from the past. Major climatic downturns such as the '536 event' with marked impact in several parts of the globe from 536 CE onwards until the middle of the century (Newfield 2018) have left plenty of evidence in the archaeological record around the world.

A common trait of the archaeological record and the impacts of the living and non-living material world is that they are not exclusive to humans. The fact that the archaeological record draws a picture of the past as a complex and intertwined fabric of human and more-than-human elements rather than a mere spectacle of past humanity is increasingly acknowledged as a premise of archaeological theory (e.g., Fredengren (2021), cf. Bradley (2021), Cipolla et al. (2021)).

Overton and Hamilakis (2013) argue for a shift from zooarchaeology to social zooarchaeology, underlining how archaeological study of animal remains can contribute to the understanding of not only the human but also the animal past. Lucas (2018) calls attention to what he terms symbiotic architectures that are animate assemblages of human and nonhuman elements. Similarly to how the example of a turf house distorts the lines between the past and the present, focusing on the living, Stobiecka's (2022) notion of lively heritage focuses attention on how the more-than-human entanglements of the present contribute to the understanding of what can be described as its *longue durée*.

Archaeological Sites as Multispecies Information Habitats

In parallel to offering new perspectives to enrich archaeological inquiry and understanding of the past, the broadening attention to the more-than-human nature of the archaeological record has underscored its usefulness in cross-disciplinary undertakings. Archaeology is increasingly entwined in the multidisciplinary research contributing to addressing global challenges, including climate change and sustainability (e.g., Huvila, Dallas et al. 2022; Schofield and Schofield 2024). In parallel, the post-human and more-than-human theorising informs archaeological inquiry (Cipolla et al. 2021), raising questions of the relation of the discipline to new uses of the archaeological record and its positionality in relation to multispecies perspectives. Similarly to how posthumanisms of the present question the human privilege in the contemporary setting, more-than-human archaeologies do it for the past.

Multispecies Information Practices of the Past

However, rather than enquiring into how archaeology has engaged in posthumanist theory and practice (Cipolla et al. 2021), from the perspective of this chapter with an aim of showing how archaeological practice can provide stratagems for approaching contemporary multispecies information practice, it is more interesting to focus on how archaeology approaches archaeological sites as what can be termed *multispecies information habitats* or, in other words, living ecosystems of information making and taking that do not *a priori* privilege human agency and involvement.

Even if the focus of archaeological enquiry is on past human presence, it unfolds through human engagements with human and the nonhuman others. Even in its criticised anthropocentrism, archaeologists' approach that takes the nonhuman world seriously as a source of information about past human undertakings is something information science could find useful. The same applies undoubtedly to the archaeological interest in Indigenous knowledge practices as a means to overcome the anthropocentrism of the predominant Western ontologies (e.g., Lane 2021; Motta and Porr 2023; Stobiecka 2022;

Wallis 2023), however, as for archaeologists, not only as fixed alternatives to consider but also as reminders of the close-to-endless possibilities of how multispecies information habitats might be configured in the most unexpected manners.

Present Information Practices of the Multispecies Past

Parts of the archaeological record and multidisciplinarity give more prominence to nonhuman aspects and perspectives than others. Osteology, archaeobotany and soil analysis focus to a large extent on non-artefactual and nonhuman materials. Environmental archaeology, but also many other archaeological research fields, have to varying degrees been more-than-human long before the introduction of contemporary archaeological posthumanisms (Cipolla et al. 2021) that draw on interdisciplinary theorising. However, the focus of inquiry does not equate with its perspective. Investigating materials or phenomena of nonhuman origins can still focus very much on human engagements, similarly to how human evidence can be approached to shed light on the more-than-human past.

Overton and Hamilakis' (2013) critique of the anthropocentricity of conventional zooarchaeology strikes into this distinction. Their thesis for social zooarchaeology is to establish it as a form of multispecies archaeology that acknowledges the presence and possibility of multiple alternative ontologies and the constitution of the world not as a human but as a co-constituted achievement. At the same time, they and later multispecies archaeologists alike emphasise how multispecies archaeological research benefits greatly from the work that does not share its pluralistic theoretical ambitions. The archaeological imperative to work on whatever material is available is helpful in avoiding losing sight of these opportunities. As Heringman (2018) notes, how early antiquarianism used travel accounts to illustrate volcanic shaping of the landscape, a piece of archaeological evidence can be helpful even if it has not been collected or generated from a more-than-human perspective or if it only indirectly evinces the agency of nonhumans.

Limits of Archaeological Engagements

While emphasising how archaeology is successful in generating knowledge of extremely heterogeneous and fragmentary more-than-human assemblage things, it is important to remember its limitations. Even if archaeology has, in parallel to human sciences in general, engaged with its own subjectivity with key works published in relation to decolonial (e.g., Smith and Wobst 2004; Hamilakis 2018; Lane 2021), feminist (e.g., Gero 1996; Conkey and Gero 1997) and reflexive (e.g., Hodder 2000; Berggren 2014) archaeologies, and the studies of archaeological practices have underlined its bodily dimension (e.g., Goodwin 1994;

Olsson 2016), there are limits to the extent archaeology has reflected its relation to embodiment and 'existential condition' (Csordas 2025) in relation to the world. The same applies to how mainstream archaeology still often approaches, describes and problematises multispecies interactions in the archaeological record (Motta and Porr, 2023, with multiple exceptions, e.g., Overton and Hamilakis, 2013; Oma and Goldhahn, 2020) in a sense that significantly differs from how, for example, some of the recent work in conversation analysis, linguistics, and biosemiotics has engaged in comprehensively rethinking their research practice in more-than-human terms (cf., e.g., Peltola and Simonen 2024). An even greater challenge for archaeology is, however, that many aspects of the past are bound to remain a conundrum even if methodological and theoretical developments in the field are constantly leading to new breakthroughs. Unlike in studies focusing on the present, like much of information science research, no new evidence of the past can be generated at the discretion of the researchers. Archaeology, as any field of research, is also limited from its outset by its methods and theories and the context within which it operates (e.g., Leighton 2015; Huggett 2022). While archaeological inquiry opens black boxes, when they are opened, there are still many hollows to fill.

Archaeological Perspectives on Contemporary Information Habitats

Similarly, to the archaeological past, contemporary human habitats are not solely human spaces but also territories and ecosystems inhabited by multiple living and non-living agencies and species. Considering what archaeology does, there are different ways in which the archaeological gaze can contribute to investigating the present. In general, historical archaeology and the archaeologies of the recent past have demonstrated how archaeological inquiry complements knowledge acquired from documentary sources (Andrén 1998; Gonzalez-Ruibal 2018). More specifically for information research, I posit that archaeological gaze comes with affordances for advancing more-than-human perspectives in the study of information habitats.

The focus of archaeology in many respects, even without underlining it (Bradley 2021) as a more-than-human discipline and the archaeological record as a more-than-human record of the more-than-human past, points to the opportunities and the *critical need to problematise the dominant conceptualisations of information and information behaviour* often portrayed as all very human concepts. This applies to archetypal information—things like books or social media posts—but also to the contemporary understandings of data as a particular type of artefact with specific affordances beyond being a variety of information. The conventional thinking about information does not mean, however, that the notions of information or information behaviour—even information practices—would need to be specifically human. While information scientists like Bates (2006, 2026) and documentalists, following

Briet's (1951) classic example of an antelope as a (potential) document, have acknowledged its broader scope (more examples in Solhjoo 2025; Solhjoo and Hartel 2026), information research seldom focuses on the informationality of the more-than-human sphere and nonhuman informational agency as more than a curiosity. The tendency to focus on archetypal information-things has parallels with prehistoric archaeologists' simultaneous longing for complementary written evidence for critique of historical archaeology on giving too much precedence to it (Andrén 1998, pp. 3–4). The possibility to store and transmit information in written form makes reality 'stable, mobile, comparable and combinable' (Miller and Rose 2008, p. 65), enabling what Latour terms to 'act at a distance' (Latour 1987, p. 223; Shove 2022), but at the same time, it takes over the practice of everyday life. The (non-written) archaeological record is—precisely because of this contradiction—a useful point of comparison to the information researchers' 'contemporary record' in how it can help to broaden the vista of what can be, and regularly is, 'taken' as information.

A remark of Lane on the popular misconception that the idea of studying material remains to enquire into the past is a Western construct provides another interesting point of comparison to the similarly customary assumptions of what information is and for whom. Compared to how, according to Lane (2021), archaeology could be a human universal, information could be and would benefit from being approached more often as a more-than-human universal. The crux when doing this, however, is, as the variety of ways of how the term 'information' is used across diverse academic and non-academic contexts, to strike into this diversity and, as Lane (2021) urges archaeologists to do, to focus on affinities instead of contrasts between knowledge systems. For multispecies more-than-human information, this could mean putting more focus on diverse informationalities in things, processes and ontologies: What is being made and taken, and how, and where all of it comes together.

Multispecies archaeologies also evince the *difficulty, special adeptness, competence and affinity needed to be attentive to the nonhuman sphere* in order to incorporate it in archaeological research. This resonates with what has been experienced in multispecies information research (e.g., Sohljoo, in this volume). Many archaeologists and anthropologists also argue that understanding animistic and relational ontologies requires subjective experience of them (Carocci 2023). For instance, Goldhahn (2023) notes how several notable archaeologists of bird remains were also renowned ornithologists. Olsson and Lloyd's account of a geologist on an archaeological site provides another example. The nonhuman becomes meaningful only when the spectator knows what to look for and how. However, at the same time, archaeology is empathetically an endeavour of enquiring into the other. It is making sense of the past, which is, as Hartley (1953) famously phrased and Lowenthal (1985) popularised, a foreign country, or even more strange and difficult to approach than that. In this sense, attempts to approach the multispecies other at the present can learn from archaeological thinking. An archaeologist needs to be careful in avoiding rigid preconceptions of how specific kinds of artefacts might have been used or how they might have ended up being discovered in particular places (Hodder 2012; Aldred 2020). At the same time, as the long-running debate in archaeology evinces (e.g., Ascher 1961; Wylie 1985; Spriggs 2008), making comparisons and analogies between how different individuals, groups or species act can be as helpful as it can be risky.

Another benefit of thinking informationally through the lens of archaeological practice is in how it can provide stratagems for approaching contemporary multispecies information habitats from the outset of how an inquiry into nonhuman entities contributes to understanding past knowledge-making activities. As the theory of making and taking suggests, *there is much more to be taken as information as information research typically acknowledges*—from artefacts to natural materials (Huvila 2014)—and approaches to engage with them through, for example, listening (Supper and Bijsterveld 2015; Stewart-Robertson 2024), smelling (Zuboff 1988), feeling, and tasting (e.g., Olsson 2016; Huvila 2019b) them. The often taken-for-granted notions of information that risk impoverishing the understanding of the engagements in the more-than-human pluriverse (Cadena and Escobar 2023) can also fruitfully be problematised from the outset by the diversity of what archaeological practice considers as informative. An important insight is also how different types of information work together. Feeling or looking at a physical object or considering archaeological animal remains as evidence of utilisation or co-presence are not separate undertakings. Similarly, one piece of information can be informative to multiple species, and seldom is anyone working with only one source of information at a time. Further, the empirical openings also have consequences for the extent to which information is an 'evolutionary instinct' (Spink 2010) specific to particular species and, for example, how inherently documents and documenting are tied to humans and their expressions (cf. Lund 2024; Olsen-Kristiansen and Lund 2024). Rather than assuming that any 'information' and 'document,' independent of how they are defined, have to be only human, or suggesting that anything can be information and any living being is working with documents, following archaeologists' pursuits to understand the past can help to keep them as open questions also in contexts where such questions might unfold as uncomfortably settled and unproblematic from the outset.

Archaeological practice can also help to *imagine interconnectivities of things*. Even if the selling point of information research reasonably is its focus on following the red thread of information, there has been a relative dearth, especially of theorising, of the nexus of information and practices. As a concept, information is a useful abstraction, but failing to account for the variety of ways in which the world comes together can turn it into a hindrance to capturing the diversity of what Abram (1997) terms the sensuous world. Similarly to how having a hammer in a hand makes us see things as nails, an information perspective easily turns an information scientist into seeing all problems as information problems. Archaeologies of multispecies information habitats can lead us to make sense of the mechanisms of how information is entangled in the more-than-human existence, for example, how the authoring of archaeological documentation authors archaeological practice and engagements with the past (cf. Huvila 2019a), and in the broader sense, how information ties into other facets of designing and imagining ethical and sustainable strategies for future cohabitation of humans and nonhumans in more-than-human worlds. A more environmentally and, in parallel and simultaneously, epistemically and ontologically sustainable cohabitation requires what Varutti (2025) terms empathetic attention, a feeling of kinship and its enactment in practices that are ecological in an ecosystemic sense. Such designs have rightly been pointed out to be more important than ever to overcome limitations of human-centrism (Wakkary 2021) and require knowledge of both human and nonhuman spheres (Roudavski 2020); however, rather than as two separate entities, they should be approached as an inseparable territory. Similarly to how environment and technology come together in envirotechnological systems, information is intertwined rather than merged in information ecosystems that are, similarly to envirotechnological systems, 'both human and nonhuman, artificial and natural, technological and ecological' (LeCain 2009, p. 22). Such endeavours benefit from knowledge of past successes and failures—how more-than-human engagements of the past have prospered or disappeared.

Finally, engaging with archaeological practice also makes apparent how the *relationships between living and non-living agencies differ across contexts and how they shift in time*. Information research has been struggling for a long time with the apparent contextuality of information and information practices, cross-contextual similarities, and ambitions to extrapolate information as a global good. Even if both archaeology and information science operate productively with analogies between situations and times, acknowledging Hartley's maxim that '[t]he past is a foreign country; they do things differently there' (Hartley 1953, p. 1) in practice in an archaeological sense is a useful starting point for enquiring and imagining how things can be different—even if there would be a lot of similarities—in the past but also in other presents and futures. Assuming similarity or difference, locality or globality, rather than asking how they are enacted, is equally risky. Krarup's (2024) urge to pragmatist social scientists to follow archaeologists in focusing on the necessary conditions of existence of phenomena rather than attending to (seemingly) general patterns, and to juxtapose contextually close matters rather than to foreground lookalikes, is useful advice for information scientists as well. Archaeology shows how the similar appearance of practices or physical objects does not need to suggest their common origins. Different living and non-living beings travel similarly to ideas. But all of them can also remind us of others emerging independently in multiple locations and points in time. The archaeological record also provides plentiful evidence that the relations of animals, plants, and humans have not always been the same. It evinces coexistencies that have been, locally, both more and less destructive and respectful than those at present. Considering such alternatives provides a starting point for envisioning how to take into account 'more-than-human concerns' (Akama et al. 2020) also through the information lens: How informational phenomena and practices really are related and unrelated, what consequences and opportunities diverse practices of informing and getting informed, organising, managing and keeping, using and disposing potentially informative things have and for whom.

Conclusions

In this chapter, we have taken a short tour into the multispecies more-than-human past, where human and nonhuman entities and their lives are intricately interwoven into each other. These entanglements are visible both in the practices of archaeologists and others who study the past at the present, but also in the past itself, which the archaeological research is trying to help us to understand. A glimpse into the contemporary habitats of archaeological knowledge, the ecologies of multispecies and more-than-human practices of archaeologists to make and take archaeological information to contribute to the understanding of the past, opens a broad vista to a massive and intricate informational enterprise where the multispecies and more-than-human nature of information practices is readily apparent.

In parallel, the inquiry into archaeological practice provides cues to how to approach other, both obviously and less apparently, multispecies information habitats in information science research. The diversity of things takeable as information and the variety of practices of how information can be generated and appropriated in use underline the need to stay critical of information and information behaviour as only or even primarily human concepts. The conundrums of taking information in a multispecies more-than-human world underline the need to augment the idea of what information is and when, how diverse informative things are interconnected, and how they differ across contexts and time. This also underlines the need for specific, sometimes unexpected, competences and affinities from those who are engaging in information practices.

Rather than being content with enquiring into human information practices and their information ecologies, there is room for a shift towards delving into information-in-practices (cf. Cox 2013) and information-in-ecologies that are multispecies and more-than-human. The engagements with the multispecies more-than-human past require both new concepts and a new gaze at the informational world. There is a need for what Calarco (2008) terms 'letting go' of unnecessarily rigid categories when shifting perspective to the multispecies informational habitats. As archaeologist Lucas (2018) urges his colleagues, acknowledging the agency of multiple species requires giving up agency in stories that traditionally have been seen as belonging to researchers. The same could be said of information science. Taking multispecies and more-than-human seriously requires new stories of information as more-than-human, processual and much more complex to make and take than often imagined.

References

Abram, David (1997). *The spell of the sensuous: Perception and language in a more-than-human world.* New York: Vintage.

Akama, Yoko, Light, Ann, & Kamihira, Takahito (2020). Expanding participation to design with more-than-human concerns. In *Proceedings of the 16th Participatory Design Conference 2020 - Participation(s) Otherwise - Volume 1* (pp. 1–11). Manizales Colombia: ACM. https://doi.org/10.1145/3385010.3385016

Aldred, Oscar (2020). *The archaeology of movement.* London: Routledge. https://doi.org/10.4324/9780429203022

Amsterdamska, O. (1990). Surely you are joking, monsieur latour! *Science, Technology, & Human Values, 15*(4), 495–504.

Andrén, Anders (1998). Between artifacts and texts: Historical archaeology in global perspective. In Charles E. Orser (Ed.), *Contributions to global historical archaeology.* Boston, MA: Springer. https://doi.org/10.1007/978-1-4757-9409-0

Ascher, Robert (1961). Analogy in archaeological interpretation. *Southwestern Journal of Anthropology, 17*(4), 317–325. https://doi.org/10.1086/soutjanth.17.4.3628943

Barad, Karen (2007). *Meeting the universe half-way.* Durham, NC: Duke University Press.

Bates, Marcia J. (1999). The invisible substrate of information science. *Journal of the American Society for Information Science, 50*(12), 1043–1050. https://doi.org/10.1002/(SICI)1097-4571(1999)50:12<1043::AID-ASI1>3.0.CO;2-X

Bates, Marcia J. (2006). Fundamental forms of information. *Journal of the American Society for Information Science and Technology, 57*(8), 1033–1045.

Bates, Marcia J. (2017). Information behavior. In J. D. McDonald & M. Levine-Clark (Eds.), *Encyclopedia of library and information sciences* (3rd ed.). CRC Press.

Bates, Marcia J. (2026, this volume). Researching information behaviour among multiple species. In N. Solhjoo (Ed.), *Multispecies information science*, 22–38. London: Routledge.

Berggren, Åsa (2014). Reflexive approaches in archaeology, development of. In Claire Smith (Ed.), *Encyclopedia of global archaeology* (pp. 6249–6257). Springer. https://doi.org/10.1007/978-1-4419-0465-2_1023

Birch, Suzanne E. Pilaar (2018). Introduction. In Suzanne E. Pilaar Birch (Eds.), *Multispecies archaeology* (pp. 1–7). London: Routledge. https://doi.org/10.4324/9781315707709

Bradley, Richard (2021). More than. *Current Swedish Archaeology, 29*(1), 48–51. https://doi.org/10.37718/CSA.2021.05

Briet, Suzanne (1951). *Qu'est-Ce Que La documentation?* Paris: EDIT.

Cadena, Marisol de la, & Escobar, Arturo (2023). Notes on excess: Towards pluriversal design. In Martín Tironi, Marcos Chilet, Carola Ureta Marín, & Pablo Hermansen (Eds.), *Design for more-than-human futures* (pp. 29–50). London: Routledge.

Cahoon, Neal Peter (2025). Researching "with" through practice-research methodologies. *Sosiologia, 62*(1), 20–38.

Calarco, Matthew (2008). *Zoographies: The question of the animal from Heidegger to Derrida.* New York: Columbia University Press.

Carocci, Max (2023). Raptors in Pre-Columbian North America: An ontology of art. In Robert J. Wallis (Ed.), *The art and archaeology of human engagements with birds of prey: From prehistory to the present* (pp. 136–151). New York: Bloomsbury.

Cipolla, Craig N., Crellin, Rachel J., Harris, Adrienne M., & Harris, Oliver J.T. (2021). Posthuman archaeologies, archaeological posthumanisms. *Journal of Posthumanism*, *1*(1), 5–21.

Conkey, Margaret W., & Gero, Joan M. (1997). Programme to practice: Gender and feminism in archaeology. *Annual Review of Anthropology*, *26*(1), 411–437. https://doi.org/10.1146/annurev.anthro.26.1.411

Cox, Andrew M. (2013). Information in social practice: A practice approach to understanding information activities in personal photography. *Journal of Information Science*, *39*(1), 61–72. https://doi.org/10.1177/0165551512469767

Csordas, Thomas J. (2025). Something other than its own mass: Embodiment as corporeality, animality, & materiality. *Anthropological Theory*, *25*(1), 3–29. https://doi.org/10.1177/14634996241230169

Fredengren, Christina (2021). Beyond entanglement. *Current Swedish Archaeology*, *29*(1), 11–33. https://doi.org/10.37718/CSA.2021.01

Gero, Joan (1996). Archaeological practice and gendered encounters with field data. In Rita P. Wright (Ed.), *Gender and archaeology* (pp. 251–280). Philadelphia: University of Pennsylvania Press.

Gibson, J. J. (1977). The theory of affordances. In R. Shaw & J. Bransford (Eds.), *Perceiving, acting, and knowing: Toward an ecological psychology* (pp. 67–82). Hillsdale, NJ: Lawrence Erlbaum.

Goldhahn, Joakim (2023). Ghosthawk worldings: Raptor haruspicy during the North European Bronze age. In Robert J. Wallis (Ed.), *The art and archaeology of human engagements with birds of prey: From prehistory to the present* (pp. 53–68). New York: Bloomsbury.

Gonzalez-Ruibal, Alfredo (2018). *An archaeology of the contemporary era*. London: Routledge. https://doi.org/10.4324/9780429441752

Goodwin, Charles (1994). Professional vision. *American Anthropologist*, *96*(4), 606–633.

Greifeneder, Elke, & Schlebbe, Kirsten (2022). Information behaviour. In Rainer Kuhlen, Dirk Lewandowski, Wolfgang Semar, & Christa Womser-Hacker (Eds.), *Grundlagen der informationswissenschaft* (pp. 497–510). Berlin: De Gruyter Saur.

Hamilakis, Yannis (2018). Decolonial archaeology as social justice. *Antiquity. A Quarterly Review of Archaeology*, *92*(362), 518–520. https://doi.org/10.15184/aqy.2018.17

Hamilakis, Yannis, & Overton, Nick J. (2013). A multi-species archaeology. *Archaeological Dialogues*, *20*(2), 159–173. https://doi.org/10.1017/S1380203813000214

Haraway, Donna (2003). *The companion species manifesto: Dogs, people, and significant otherness*. Chicago, IL: Prickly Paradigm Press.

Hartley, L. P. (1953). *The go-between. New York review books classics*. London: New York Review Books; Distributed by Publishers Group West.

Heringman, Noah (2018). Calabrian hounds and roasted ivory (or, swerving from anthropocentrism). In S. P. Birch (Ed.), *Multispecies archaeology* (pp. 11–25). London: Routledge.

Hodder, Ian (Ed.) (2000). *Towards reflexive method in archaeology: The example at Çatalhöyük*. Cambridge: McDonald Institute for Archaeological Research.

Hodder, Ian (Ed.) (2012). *Entangled: An archaeology of the relationships between humans and things*. Malden, MA: Wiley-Blackwell.

Huggett, Jeremy (2022). Data legacies, epistemic anxieties, and digital imaginaries in archaeology. *Digital*, *2*(2), 267–295. https://doi.org/10.3390/digital2020016

Huvila, Isto (2009). Ecological framework of information interactions and information infrastructures. *Journal of Information Science*, *35*(6), 695–708. https://doi.org/10.1177/0165551509336705

Huvila, Isto (2014). Archaeologists and their information sources. In Isto Huvila (Ed.), *Perspectives to archaeological information in the digital society* (pp. 25–54). Uppsala: Department of ALM, Uppsala University.

Huvila, Isto (2015). Situational appropriation of information. *Aslib Journal of Information Management*, *67*(5), 492–504. https://doi.org/10.1108/AJIM-02-2015-0033

Huvila, Isto (2018). Ecology of archaeological information work. In Isto Huvila (Ed.), *Archaeology and archaeological information in the digital society* (pp. 121–141). London: Routledge.

Huvila, Isto. 2019a. Authoring social reality with documents: From authorship of documents and documentary boundary objects to practical authorship. *Journal of Documentation*, *75*(1), 44–61. https://doi.org/10.1108/jd-04-2018-0063

Huvila, Isto. (2019b). Learning to work between information infrastructures. *Information Research*, *24*(2), 819.

Huvila, Isto (2019c). Rethinking context in information research: Bounded versus Centred sets. *Information Research*, *24*(4), paper colis1912.

Huvila, Isto (2019d). Genres and situational appropriation of information: Explaining not-seeking of information. *Journal of Documentation*, *75*(6), 1503–1515. https://doi.org/10.1108/jd-03-2019-0044

Huvila, Isto (2022). Making and taking information. *JASIST*, *73*(4), 528–541. https://doi.org/10.1002/asi.24599

Huvila, Isto, Dallas, Costis, Toumpouri, Marina, & Enqvist, Delia Ní Chíobháin (2022). Archaeological practices and societal challenges. *Open Archaeology*, *8*(1), 296–305. https://doi.org/10.1515/opar-2022-0242

Huvila, Isto, Enwald, Heidi, Eriksson-Backa, Kristina, Liu, Ying-Hsang, & Hirvonen, Noora (2022). Information behavior and practices research informing information systems design. *JASIST*, *73*(7), 1043–1057. https://doi.org/10.1002/asi.24611

Introne, Joshua, McKernan, Brian, Corsbie-Massay, Charisse L' Pree, Rohlinger, Deana, & Tripodi, Francesca Bolla (2024). Healthier information ecosystems: A definition and agenda. *Journal of the Association for Information Science and Technology*, *75*(10), 1025–1040. https://doi.org/10.1002/asi.24949

Johanson, Kristiina. (2018, December). Missing interpretations: Tracing natural artefacts in Estonian archaeological material. *Fennoscandia Archaeologica*, *35*, 87–106.

Killick, David (2015) Using evidence from natural sciences in archaeology. In Robert Chapman & Alison Wylie (Eds.), *Material evidence* (pp. 159–172). London: Routledge.

Krarup, Troels. (2024, October). What social research can learn from archaeology: Comparison as juxtaposition and conduction. *Theory, Culture & Society*. https://doi.org/10.1177/02632764241280216

Lane, Paul (2021). Archaeology, hybrid knowledge, & community engagement in Africa: Thoughts on decolonising practice. In A. McGrath & L. Russell (Eds.), *The Routledge companion to global indigenous history* (pp. 603–628). London: Routledge.

Latour, Bruno (1987). *Science in action: How to follow scientists and engineers through society*. Cambridge, MA: Harvard University Press.

Latour, Bruno (1988). *The pasteurization of France*. Cambridge, MA; London: Harvard University Press.

LeCain, Timothy J. (2009). *Mass destruction: The men and giant mines that wired America and scarred the planet*. New Brunswick, NJ: Rutgers University Press.

Leek, F. F. (1972). Teeth and bread in ancient egypt. *The Journal of Egyptian Archaeology*, *58*(1), 126–132. https://doi.org/10.1177/030751337205800109

Leighton, Mary (2015). Excavation methodologies and labour as epistemic concerns in the practice of archaeology. Comparing examples from British and Andean archaeology. *Archaeological Dialogues*, *22*(1), 65–88. https://doi.org/10.1017/s1380203815000100

Lowenthal, David (1985). *The past is a foreign country*. Cambridge: Cambridge University Press.

Lucas, Gavin (2012). *Understanding the archaeological record*. Cambridge: Cambridge University Press.

Lucas, Gavin (2018). Symbiotic architectures. In Suzanne E. Pilaar Birch (Ed.), *Multispecies archaeology* (1st ed., pp. 105–117). Routledge. https://doi.org/10.4324/9781315707709-7

Lund, Niels Windfeld (2024). *Introduction to documentation studies*. London: Facet.

Miller, Peter, & Rose, Nikolas S. (2008). *Governing the present: Administering economic, social and personal life*. Cambridge: Polity.

Morton, Timothy (2010). *The ecological thought*. Cambridge, MA: Harvard University Press. https://doi.org/10.2307/j.ctvjhzskj

Motta, Ana Paula, & Porr, Martin (2023). The Jaguar Gaze: Is it possible to Decolonise human–animal relationships through archaeology? In Rick De Vos (Ed.), *Decolonising animals* (pp. 185–218). Sydney University Press. https://doi.org/10.2307/jj.1094298.11

Newfield, Timothy P. (2018). The climate downturn of 536–50. In Sam White, Christian Pfister, & Franz Mauelshagen (Eds.), *The Palgrave Handbook of climate history* (pp. 447–493). London: Palgrave Macmillan UK. https://doi.org/10.1057/978-1-137-43020-5_32

Olsen, Bjørnar (2012). *Archaeology the discipline of things*. Berkeley: University of California Press.

Olsen-Kristiansen, Bernt Ivar, & Lund, Niels (2024). Documents in the age of non-human agency. *Proceedings from the Document Academy*, *11*(2). https://doi.org/10.35492/docam/112/10

Olsson, Michael (2016). Making sense of the past: The embodied information practices of field archaeologists. *Journal of Information Science*, *42*(3), 410–419. https://doi.org/10.1177/0165551515621839

Olsson, Michael, & Lloyd, Annemaree (2017). Being in place: Embodied information practices. *Information Research*, *22*(1), 1601.

Oma, Kristin Armstrong, & Goldhahn, Joakim (2020). Human-animal relationships from a long-term perspective. *Current Swedish Archaeology*, *28*(1), 11–22. https://doi.org/10.37718/CSA.2020.01

Overton, Nick J., & Hamilakis, Yannis (2013). A manifesto for a social zooarchaeology. Swans and other beings in the mesolithic. *Archaeological Dialogues*, *20*(2), 111–136. https://doi.org/10.1017/S1380203813000159

Patrik, Linda (1985). Is there an archaeological record? *Advances in Archaeological Method and Theory*, *8*, 27–62.

Peltola, Rea, & Simonen, Mika (2024, January). Towards interspecies pragmatics: Language use and embodied interaction in human-animal activities, encounters, and narratives. *Journal of Pragmatics*, *220*, 15–19. https://doi.org/10.1016/j.pragma.2023.11.013

Pickering, Andrew (1995). *The mangle of practice: Time, agency, and science*. Chicago, IL: University of Chicago Press.

Pitblado, Bonnie L., Thomas, Suzie, Wessman, Anna, & Woodward, Sophie (2025, February). Retheorizing archaeological "artefacts" as "belongings". *Archaeologies*. https://doi.org/10.1007/s11759-025-09523-1

Pritchard, Sara B. (2011). *Confluence: The nature of technology and the remaking of the Rhône*. Cambridge, MA: Harvard University Press.

Pye, Jeremy W. (2021). 'Unwanted guests': Evidence of parasitic infections in archaeological mortuary contexts. *Historical Archaeology*, *55*(1), 82–95. https://doi.org/10.1007/s41636-020-00271-3

Roudavski, Stanislav (2020). Multispecies cohabitation and future design. In *Design Research Society Conference 2020*. https://doi.org/10.21606/drs.2020.402

Schofield, John, & Schofield, John (2024). *Wicked problems for archaeologists: Heritage as transformative practice*. Oxford: Oxford University Press.

Shove, Elizabeth (2022). *Connecting practices: Large topics in society and social theory*. London: Routledge. https://doi.org/10.4324/9781003275107

Smith, Claire, & Martin Wobst, H. (Eds.) (2004). *Indigenous archaeologies: Decolonising theory and practice*. London: Routledge. https://doi.org/10.4324/9780203009895

Solhjoo, Niloofar (2025). The multispecies perspective in library and information science. *Information Research an International Electronic Journal*, *30*(CoLIS), 323–330. https://doi.org/10.47989/ir30CoLIS51898

Solhjoo, Niloofar, & Hartel, Jenna (2026). The footprints of animals in library and information science. In N. Solhjoo (Ed.), *Multispecies information science*, 5–21. London: Routledge.

Spink, Amanda (2010). *Information behavior: An evolutionary instinct*. Berlin – New York: Springer.

Spriggs, Matthew (2008). Ethnographic parallels and the denial of history. *World Archaeology*, *40*(4), 538–552. https://doi.org/10.1080/00438240802453161

Stengers, Isabelle (2005). Introductory notes on an ecology of practices. *Cultural Studies Review*, *11*(1), 183–196.

Stewart-Robertson, Owen (2024). Information practices and the production of scientific knowledge through sound. *Proceedings of the Association for Information Science and Technology*, *61*(1), 1086–1089. https://doi.org/10.1002/pra2.1193

Stobiecka, Monika (2022). Lively heritage: On more-than-human encounters at Mediterranean archaeological sites. *Journal of Contemporary Archaeology*, *9*(1), 64–81. https://doi.org/10.1558/jca.21635

Supper, Alexandra, & Bijsterveld, Karin (2015). Sounds convincing: Modes of listening and sonic skills in knowledge making. *Interdisciplinary Science Reviews*, *40*(2), 124–144. https://doi.org/10.1179/0308018815Z.000000000109

Tsing, Anna Lowenhaupt, Bubandt, Nils, Gan, Elaine, & Swanson, Heather Anne (2017). *Arts of living on a damaged planet: Ghosts and monsters of the anthropocene*. Minneapolis: University of Minnesota Press.

Varutti, Marzia (2025). (Re-)learning to relate to the more-than-human: Some conceptual, affective, and practice-based perspectives. *Suomen Antropologi: Journal of the Finnish Anthropological Society*, *49*(2), 41–59. https://doi.org/10.30676/jfas.141513

Wakkary, Ron (2021). *Things we could design: For more than human-centered worlds*. Cambridge, MA: MIT Press.

Wallis, Robert J. (Ed.) (2023). *The art and archaeology of human engagements with birds of prey: From prehistory to the present*. New York: Bloomsbury.

Whitehead, Alfred North (1978). *Process and reality – An essay in cosmology* (2nd ed.). New York: Free Press.

Wilhelmson, Helene, & Dell'Unto, Nicolò (2015). Virtual taphonomy: A new method integrating excavation and postprocessing in an archaeological context. *American Journal of Physics Anthropology*, *157*(2). https://doi.org/10.1002/ajpa.22715

Wilhelmson, Helene, & Douglas Price, T. (2017, April). Migration and integration on the Baltic Island of Öland in the iron age. *Journal of Archaeological Science: Reports*, *12*, 183–196. https://doi.org/10.1016/j.jasrep.2017.01.031

Winner, Langdon (1993). Upon opening the black box and finding it empty: Social constructivism and the philosophy of technology. *Science, Technology, & Human Values*, *18*(3), 362–378. https://doi.org/10.1177/016224399301800306
Woźny, Jacek (2021). Archeology as a metaphor in contemporary culture. *Qualitative Sociology Review*, *17*(1), 2–38. https://doi.org/10.18778/1733-8077.17.1.3
Wylie, Alison (1985). The reaction against analogy. In Michael B. Schiffer (Ed.), *Advances in archaeological method and theory* (pp. 63–111). Academic Press. https://doi.org/10.1016/B978-0-12-003108-5.50008-7
Zuboff, Shoshana (1988). *In the age of the smart machine. The future of work and power*. New York: Basic Books.

14

MASSIVE UNSEEN COMPANIONS AND THE DOCUMENTARY PROJECT

Erik Radio

Introduction

Within the constellation Scorpio, several thousand light-years away, is the binary star system GRO J1655−40 (ESA et al., 2002). The system is about 800 million years old and consists of two celestial bodies: An old star and a black hole that accretes matter from its companion (Abramowicz and Kluźniak, 2001). Our knowledge of the existence of the black hole depends on its relation to other celestial bodies. Without the matter trailing off the GRO J1655−40 star, we might still be able to deduce the existence of the black hole due to the former's frequent orbiting of what appears as empty space. Due to how one is identified by the other, it is tempting from a classificatory perspective to identify the star with the name GRO J1655−40 and assign the black hole some other designation. In lieu of a name, one term used to describe the black hole and similar celestial phenomena is a 'massive unseen companion' (MUC).

On Earth, a similar phenomenon has emerged. As the Anthropocene advances, the documentary project—the production and dissemination of documents towards historicising ends—finds itself between several concurrent tensions. Important among these is a mass extinction event, which, along with the devastating loss of countless species, collaterally consumes unknown types of multispecies knowledge (Cowie et al., 2022). How humans are to record the loss of the multispecial[1] at a time of great ecological upheaval requires that any documentation of extinction avoid becoming purely reactionary. In other words, it is important to find ways to document extinction as an intervention and not just as a record of events and absences. To do so requires an examination into three areas. First, a dialectical analysis of the complex terrain of absence in order to understand how it unfolds in the extinction of various

DOI: 10.4324/9781003583424-16

species and how the motif of the MUC can serve as a reflection point for these concerns. Second, an analysis of the materiality of information as it pertains to nonhuman documents to illuminate the scope of human documentary projects. On these bases, it is possible to reconsider extinction through the lens of documentary inscription and signification, with a particular focus on how the absence of the multispecial, as a type of MUC, can drive multispecies investigations. Through these directions, this chapter proposes a path forward for thinking through the unfolding tension between document signification and the spectres of the multispecial.

Two Storms

As an unprecedented geological event, the Anthropocene necessitates a reconsideration of how we understand humans and their relation to other species and the planet as a whole. Documentation, as a fundamental, though not uniquely human act, must likewise be evaluated differently, especially as it concerns its ability to record this event and for perpetuity. In a previous article, I examined how framing the Earth as a document and the Anthropocene as a type of inscription allowed us to reframe documents and their legibility for beings in the post-Anthropocene (Radio, 2020). That examination took a much longer view, and so it is also important that we address questions of documentation's role in the short term. If we accept that the Anthropocene represents a 'wrong state of things' (Adorno, 2015), then how we use documents as a material intervention becomes relevant, and it is necessary to ask. The extinction of various species is one of the more immediate and jarring outcomes of the Anthropocene and one of the most complex to document, as it concerns absence. My argument in this section consists of two interlocking approaches: First, that attempting to abstract out an approach to documentation that does not address the concrete will necessarily run into difficulties. Second, the MUC, introduced above, offers a useful material framework to understand absence in a multispecial context.

Absence initially appears as a relatively simple concept. But on even a slight examination, a whole dimension of possibility appears. Absence can be expected or unexpected. It can be unknown, as when an unseen cloud evaporates. Absence is also closely linked with knowledge and, consequently, ignorance. In this context, Croissant's (2014) explication of agnotology, or the study of ignorance, highlights several more gradations of absence: Privations, silences, invisibilities, and others. In a similar way, Bernstein (2009) finds an antinomical symmetry between knowledges and non-knowledges (e.g., facts/errors, reason/folly). His efforts are rooted in an impulse towards a taxonomising of non-knowledges in order to allow them to be incorporated into the traditional library organisation of resources. Following this view, it is possible to use the classification of absence as a means of theory building. But in both cases, it should be asked whether organising absence is the best way to understand it.

As a definitional approach, taxonomy encounters several challenges. It is in line with what Adorno called 'identity thinking,' or knowing an object as the sum of its classifications (Jarvis, 1998). Such a process inevitably forecloses knowledge of an object as it is subsumed under its concept. So instead of the concept illuminating the object, the object becomes an example of the concept (Adorno, 2015). For example, saying that 'extinction is a type of hiddenness' obviously leaves out other types of hiddenness and characteristics that would not apply to extinction. By taking this route, we would inevitably fall into difficulties if we were to use it as a framwork for understanding the absence of various species. In order to determine the conditions of absence by extinction, it is necessary to step back and analyse the nature of the Anthropocene in its various manifestations. As Adorno (2015) explains of his dialectical approach, we need to '...grasp the particular not by resorting to classification but rather by disclosing its own specific character...and thus reveal the opposition by particular and universal in the object of thought in each case.' In other words, by avoiding definition and letting the concept unfold through its own actions, we can gain a greater understanding of what precisely it entails.

Considering absence within the context of the Anthropocene is no small task, as the many impacts of this event are as ephemeral as they are persistent. One framing of climate change comes from Morton (2013), who characterises it as a hyperobject—'things... massively distributed in time and space relative to humans.' But as is the case with many conclusions of New Materialisms, the insistence on the primacy of the object serves in effect to divorce the human element from the equation (Malm, 2018). Even Morton's example of the black hole as a hyperobject, something that humans of course did not create, is one that is heavily mediated due to its raw alienness. The hyperobject, which is itself both non-local yet massively distributed, does not position us well to follow the concrete movements of climate change into a deeper understanding of absence. A hurricane, which I would argue fits Morton's definition, certainly would not feel non-local to one experiencing it.

The Anthropocene should not be considered monolithically. Instead, it may be more illuminating to understand it as something having modalities. In this way, climate change can be understood as a particular form of the Anthropocene and surely its primary one. Malm (2018) has described climate change as a storm. Indeed, its effects appear to us most strikingly through the intensification of weather phenomena and dynamic temperature fluctuations. But there is the other, quieter, element in the slow persistence of rising sea levels, of winters getting harsher and summers boiling for longer. As such, the 'good' weather, the absence of the storm, may itself be as catastrophic: The sunny day that is nevertheless unseasonably warm. This small shift in thinking about the Anthropocene as a storm moves it away from its identification as a pure tempest towards encompassing all variances in weather patterns. Even in the blast of a hurricane, the centre is usually calm.

If climate change is a storm, albeit one that does not rage in a universally distributed way, then the question of its functioning as a cause of extinction needs examination. Sartre (1943) argued that storms do not actually destroy; they only modify: 'There is no *less* after a storm than before. There is only something else.' In his view, humans are the only ones capable of actual destruction. It is the element of witness, which for him is innately tied to being, that allows for destruction to be experienced as such rather than as modification. For instance, cities are destroyed and not just modified by a hurricane precisely because protections have been put up around them. But how should one think through the extinction of another species? Certainly, there are measures in place, such as preservation efforts and other protections, to avoid species eradication, but what of the unknown species that disappear before we are even able to know of their existence? Following Sartre's argument, it would appear that in the case of background extinction (extinction not caused by humans), the latter are simply modified into something else, but that raises an uncomfortable tension that appears to elevate those species which are known to exist above those that are unknown. One can be destroyed, but the other is merely turned into something else. Furthermore, it forecloses engaging with loss by reframing it as a modification. This does not yet provide us with a defensible position to understand the absences caused by the Anthropocene.

The tension between humans and storms, as a site of non-dominance, is consonant with the split between nature and society, which is itself one of the many originating strands of the Anthropocene. Dispensing with that artificial rift allows us to see that, as humans, the originators of climate change, we are intimately bound up with it to the point that we are no longer able to cleanly differentiate ourselves from it. The continued burning of fossil fuels deepens the climate crisis while requiring humans to derive ever more sources of energy to mitigate its worst effects. Humans have never been able to control the weather, but we are now finding it increasingly harder to untangle ourselves from its effects. In this way, the distinction between destruction and modification appears to dissolve.

It should be noted that we are losing some accuracy when we use 'humans' universally—a good example of the concept getting in the way of concrete specificities. While it is true that climate change originates in the human species, it is really only a small number of humans who brought fossil fuel extraction and consumption into a centralised site of persistence in the global economy. Most humans are downstream of these decisions, and Malm (2016) notes that about ⅙ of the human population should not even be considered as producing emissions. Yet it is perhaps admissible to continue to use humans heuristically as a unified category even if we briefly lose sight of this complexity. In a similar way, to talk about extinction in a totalising manner may mislead us away from the possibility that some members of a species still exist or even have adapted or evolved to a point that they are no longer recognisable. The proof of extinction is a bit of a paradox. While some species are gone, they may have evolutionary descendants that remain. The whiplash between the whole and the particular is surely felt here, but it is important to differentiate anthropic extinction from background extinction, in which the latter happened on a much longer time scale, in contrast to our current mass extinction, which is occurring in almost unmeasurably small units of geologic time.

Negation and Extinction

Following Sartre's observation that storms only modify nature, we can note that when one species disappears, the space left by it is taken up by something else. Their physical remains are decomposed or ingested by others. But nature's plenitude walks us into a contradiction when considering absence. If there is no absence, only the fullness of nature in constant flux, then there can only be positive variances and differences. Nonetheless, as already indicated above, it seems not quite right that we can move on from extinction as just another difference in the makeup of being. It is in this binary of identity/difference that the importance of negation emerges. As Jameson (2020, p. 116) explains, 'Identity necessarily produces Difference out of itself, and Difference necessarily turns back into Identity the longer we stare at it. Negativity as such emerges from the metamorphoses of the categories themselves…'

Negation, thus, is central to the understanding of extinction I am positing. As mentioned above, classificatory thought inevitably leaves something out of what it classifies, which often entangles us in a referential web whereby we are caught definitionally between concepts. In order to truly grapple with extinction, it is necessary to disclose, following Adorno's dialectical method (2015, p. 38), '… its own specific character by trying to break it open, as it were, and thus reveal the opposition…,' in other words, its negation. The concept of nature as fullness, then, must be measured against itself in order to drive it towards new understanding in the light of the absence of the multispecial. In this way, extinction emerges as its negation and opens up a different set of considerations.

Jameson (2020) has examined Deleuze's argument that negation does not appear in nature, insisting instead on existence as sheer being absent from oppositions. Negation, then, appears as anthropically imposed on nature from the outside, which requires us to ignore that humans and nature are of the same substance. By contrast, Sartre (1960) identifies need emerging from scarcity as the original negation linking humans in a relation of univocal immanence with their surroundings. One could expand this argument, noting that need is, of course, not unique to just humans but forms the basis of movement for all organic life. We can see then the importance of negation on a fundamental level as advancing concepts toward new understandings and material reality towards different ends. As it relates to extinction, difference changes the ontological footing in a way that avoids the oppositional clash that negation invites and for which it is necessary to ground it in praxis (Jameson, 2020). I will examine below how this grounding is needed to orient documents in time and move them forward out of inertia.

For now, I want to briefly focus on how the eradication of a species contains a contradiction in its own fabric. A species is eliminated, but its removal creates a new presence. Giovannoni and Quattrone (2018), following Lefebvre, argue that neither presence nor absence ever reaches an absolute state. In the case of extinction, the removal of a species from a particular ecosystem warps it as a result. For example, the loss of a predator allows other species to proliferate. These changes can introduce seismic changes to an environment, in turn intensifying the ecological impacts in a feedback loop. Extinction is not eradication or a simple subtraction. Rather, Anthropocene extinction creates a tremor of instability throughout material existence, leading to even greater crises. In this sense, extinction is not just death but a material force of disintegration.[2] To return to the anthropic storm, we could say that these species are no longer impacted by it, but their extinction intensifies it just as they are dissolved into its path.

While this is not an exhaustive examination of extinction through its negation, even this slight scan moves us away from its consideration as just a type of absence towards a different understanding, one in which it is shown to be itself a strange presence. Similar to the GRO J1655−40 black hole, these annihilated species acquire characteristics akin to a MUC: Massive in their destabilising forces and a companion in the sense of still having a material effect. Indeed, the issue between known and unknown extinction dissolves in the MUC, as both are always felt. How we think through the role of documenting such entities will require a similar examination.

Materiality and Nonhuman Information

It is clear that as both species and habitats are destroyed in the Anthropocene, so too are unknowable types and amounts of information. But is information the correct concept here? Bates (1999) examined the 'red thread of information' to demonstrate the permeability of information in people's lives, both individually and socially. Solhjoo (2024) expanded this concept into the entanglement of humans with companion animals, noting that it extends across all living beings. Bates (2026, Chapter 2) calls for extending information behaviour research beyond the human species. But when we consider this in the context of the preservation of multispecies information, we run into a material problem. As a concept, information appears almost axiomatic. To give just one example of this line of thought, Bates (2006, 2022) states that information does not consist of matter and energy but is instead the pattern of organisation of matter and energy, indicating it as some third theoretical kind.[3] Although it's a commonly held belief, the genealogy of information as an independent object is traceable and tied to the computational logic of the mid-20th century. Hayles (1999) examines the severing of information from medium in the Macy Conferences on Cybernetics, where, among other developments, she notes how Shannon and Weaver 'defined information so that it would be calculated as the same value regardless of contexts…Taken out of context, the definition allowed information to be conceptualized as if it were an entity that can flow unchanged between different material substrates….' Since information could be readily transmitted between formats, the medium became simply a carrier for information.

All of this accords with a conventional view of human documents and information, but when considering the multispecies, it becomes haunted by the same materiality from which information was supposedly divorced. Take, as an example, the city park. When dogs visit these open spaces, they usually urinate as they peruse the area, in addition to smelling the urine of the other dogs and the scents of other animals as well. Horowitz (2010) has compared this to a type of community message board. The scent of the urine is bound up with that of the soil and grass. Urine is inscribed and read. In this way, it is not unlike a document and tracks with Frohmann's (2013) argument that documentary agency does not require humans. But is there information here? Rather, is there a third object between the park and the dog that could just as easily be transmitted to a different form? Can we reinscribe the urine elsewhere and create the same information? For some theorists, the park would constitute a document as it is being used or processed for informational purposes (Buckland, 1991). However, an informational purpose would seem to amount to any experience of sensory input. This concern is enlarged when considering how much we do not know about multispecial communication and knowledge. The dog was an easy example, considering how closely tied they are to humans. How would information as a theoretical kind fare in the context of those animals that are farther removed from us evolutionarily and culturally, for which a document may not be something we can even identify, let alone split into information and carrier?

The question of how we understand multispecies information hinges on the materiality of information. Or rather, information's disembodiment is called into question when we move past the anthropic. Humans may be the species that is the best at reproducing documents, and with a velocity that gives information a reified existence as a third thing between an object and a perceiver (something like the movement in a mobius strip). But as with the park, there is no information which can be extracted, like text from an email, and placed into another medium, and provide the same signification. This McLuhanian conception of media considers the materiality of a document a critical part of the signification itself. As such, multispecies information might be more readily approached as multispecies documentation. The latter, in which the medium is forefronted, allows the material dimensions to be apprehended as the environment itself. In this light, extinction gains new complexity as environments, understood as remnant documentation, are also threatened under climate change. The park, of course, has a human-designated boundary to it, but once that is removed, so does our expectation of what a document can be for us as opposed to nonhuman organisms. As the separation of form and content disintegrates, information loses momentum as a concept and finds itself on unstable ground.

Documenting the Wrong State of Things

The trajectory taken to examine how we can understand the place of documents in a multispecies framework now leads us back to the question of how documents function in the Anthropocene. Specifically, how can the anthropic document respond to the destruction of the multispecial and multispecial documents? Documentary projects inevitably have a reactionary element, as they are necessarily after the fact. Even a document in the form of a plan is a response to a current state of affairs. But they are also more than that in being both the process and the result of the process. Similarly, GRO J1655−40 is at once the star and the light falling into its adjacent black hole. A MUC shares aspects with annihilated species in that neither is perceived but still exerts a positively felt presence. We can use this motif as the object of documentary projects, and more specifically, for how the document might be mobilised against the destruction of the multispecial.

When considering the MUC, it becomes necessary to show the warping of space and not just the regret of loss. As many species have already gone extinct due to climate change, documenting their extinction, while important, has not had the effect of halting the burning of fossil fuels. Even the prospect of our own extinction does not appear to be enough towards that end. In this way, we can understand some documents as being politically inert. Consider the classic documentary example of Briet's (2006) antelope. In her view, its presence, framed by the zoo, makes it a document. In a situation where the antelope has gone extinct, the empty cage radiates absence. The frame becomes the inscription. This is a case of the document as reaction: Its perspective looks back as it moves forward in time, like Benjamin's (2003) description of the angel of history, which looks at the past as a pile of wreckage. It thus becomes a marker of the wrong state of things. If currently the documentary project finds itself only reacting to the Anthropocene, in documenting the wrong state of things, it finds itself unable to dig itself out of a negative valence. Benjamin (2003) comes to mind again with his claim that every document of civilisation is a document of barbarism. Indeed, the fact that humans are not able to agree that the ruination of the planet is a problem, or even happening, indicates our position in a dystopian coordinate.

To counter such a programme, then, would necessitate that the document look forward to some alternative. If we are in a dystopian position, then obviously a utopian programme would be a place to look. This is not to say we need to leap ahead to a prefigured vision of the world in which the current concerns have found themselves resolved, but to recall the element of result that is processual. In other words, an alternative to the Anthropocene will not emerge without a vision, but it is in the process of providing alternatives that the document can move forward. As Jameson (2020, p. 361) notes,

> ...The most fruitful way of approaching a Utopian text or project lies not in judging its positive elements, its overt representations, but rather in seeking to grasp what it cannot (yet) think, what lies in it beyond the very limits of its own social system and of the empirical being it seeks to transcend.

It is important here to note the risk of nostalgia as a pathway when considering alternate futures. By this, I mean that there is almost always some impulse to rebuild or return to the past as a remedy for a current crisis. As Žižek (1999) explains, a current failing is not fixed by returning to the past but instead in the identification that 'the moment of the Fall...coincides with its exact opposite, with the longed-for next step in evolution.' A similar motif can be found with the MUC. As Giovannoni and Quattrone (2018, p. 854) note, 'As soon as representations attempt to fill the void, they reveal the absence of what they try to represent, and therefore make presence escape.' Placing a statue of an antelope in Briet's zoo, or a de-extincted one, would only serve to sharpen the void. But it is here that we feel the aftershock of absence, the warping of space, that serves as a different type of presence. Here we are no longer representing the loss but the effect of loss. Through this, it becomes possible to show that space, once warped, can be warped again but with a positive valence. In this way, we can see utopia as containing its own event horizon, as something receding and only ever approached, but which, without movement, is not possible.

Inscription and Intervention

While utopia provides a useful, if distant, coordinate for documentary projects, we need to return to the immediate concerns of the document in the current crisis. It should be noted that the risk of a negatively charged valence for the document can emerge. That is, instead of orienting it towards a utopian vision, its energies might collapse on itself, as a distant prospect of a better future cannot withstand the mounting challenges of cascading crises caused by climate change. Also, a possibility is that representing the warping caused by an MUC could act as a kind of blueprint. This could escalate the Anthropocene as a project by emboldening those figures for whom the domination of nature as a colonial project is a desirable opportunity. This is to say that even if a document is created as a record of the Anthropocene and not as an endorsement, it can nevertheless be vulnerable to antithetical political ends. In this way, we must address the question of inscription and whether the document, apart from utopian positioning, can, in the shorter view, become inflamed beyond co-option.

The tension between memory and material lies directly at the centre of the documentary project in the politically chaotic Anthropocene. One wants to use documents to change the state of things, but their existence can lead to a deadening effect. Adorno (2020) described the process of setting up an archive as that of turning experience into furniture, a concretisation that negates vitality. A similar observation is made by Briet that documents can obfuscate that which they represent, which Day (2026) in Chapter 1 (see Footnote 4) of this volume ties to Heidegger's thought on Being and concealment.[4] Indeed there is the real problem that documenting extinction can rob the MUC of some potentiality and, as noted, can even serve towards unintended ends. To that, we must ask if the document, with its necessity as a record, is actually the problem.

Inscription, and here I use the term to designate the production of documents as a sum process and not in the sense of text on paper, is itself a warping of space. An MUC is also a warping of space, but it differs in being more akin to a ripple extending through space and affecting it. We have noted that the document, as a representation of an MUC, can deaden or at least mute it. If we are to follow Adorno in his example, it would be better to abandon the documentary project entirely. Doing so would also abandon any prospect of halting the advance of climate change, as going quiet on the subject would not affect those currently profiting from fossil fuel extraction. Of course, the MUC, like the GRO J1655−40 black hole, will eventually dissipate until the effects of its absence are no longer perceptible or so melded into the new state of ecological affairs as to be unexaminable. The question then turns to how the document can magnify the presence of the MUC while it exists without becoming its stand-in.

We have found ourselves in a sort of paralysis, an asymmetry, between the document and the multispecial, and the document with itself. On the one hand, the need to document extinction is necessary to record the event. At the same time, the document of extinction is itself caught primarily looking back at the event with little assurance that its role as a record will prevent another recurrence. To move forward, we must accept the bond between the document and the MUC as both being in states of decay. When the document comes into existence as an indexical sign for the MUC, it both replaces and mutes it, as it represents it. But as it is itself also in decay, it runs the risk of what we might call intensified negation. By this I mean that not only does it replace the sign with itself, an inadequate substitution, but it also avoids addressing its own decay and instead embraces a perceived permanence. We fall into a bad praxis using the document as a substitute for the loss of the nonhuman while accepting the false reality of the document's permanence. This intensification of claiming permanence while itself failing to meet that criterion places the document outside of indexicality towards a total absorption of the multispecial. This steers us back towards the society and nature split, a precursor to the current crises. Considering the heavy costs and environmental impacts of mining, logging, and other extractions, we can even wonder how many species go extinct so that we can document their death.

Harnessing Impermanence

The indifference to decay can be owed to what we might call the hegemony of the synchronic. The image of the archive with its materials sorted into boxes in some ways reflects what Jameson (2003) describes as the shift from a temporal to a spatial dominant. There then emerges something of a paralysis where documents, in their spatial ordering, must grapple with the temporal nature of extinction. As the document pulls the event into its boundaries, it necessarily freezes it like Adorno's furniture. This spatialism overlaps with what Hartog (2015) describes as presentism, in which '…the present, in the very moment of its occurrence, seeks to view itself as already history.' From this, we can argue that the document becomes more interested in itself as a document of extinction rather than documenting extinction. The alternative, or one of them, would be for the document to reclaim time as part of its telos in addition to leveraging its spatial position. As an MUC enacts a warping of space that eventually dissipates, so too can the document leverage temporality towards different significations.

'If you see me, then weep.' So reads one of the hunger stones in the Czech Republic. These inscriptions on the sides of rocks appear or disappear with certain environmental conditions, in this instance with the water level of a river (Henley, 2022). These inscriptions date back to the Middle Ages and are scattered throughout central Europe. Though created in response to a specific crisis at the time, the hunger stones have built into their form a temporality that, with erosion and other forms of degradation, remains applicable to our own contexts. That these documents appear to go in and out of existence, appearing only at the onset of a disaster, gives them an almost spectral quality as a harbinger. In this way, the stones, while created in response to wreckage, look ahead towards possibilities and specifically that of recurrence. Another strength of the medium, and this is speaking with the obvious affordance of history, is that the boundary between where the document starts and the environment begins is not clear or is at best debatable. With erosion and the general wearing down of time, those boundaries become even more elusive, and the document itself becomes environmental.

An echo of this type of document can be seen in J.R. Carpenter's and Tomo Kihara's (2021) *This is Not a Good Sign*, in which augmented reality is used to overlay signage into the environment. These signs contain statements such as 'These are not natural events', 'And now fire is a season', or an arrow that says 'Warmer Days', to name a few. Here, the recognisability of traffic and road signage is defamiliarised with climatic warnings; literal signs as harbingers. Whereas the hunger stones relied on the environment for inscription and mediation, in addition to operating on a vast time scale, these signs rely on the ephemerality of digital technology to impose themselves onto a mediated environment. Both are illuminative in showing different site-specific strategies for signification with a particular concern for the temporal character of documents as interventions.

Due to the effects of erosion, some of the hunger stones are no longer legible. While certain characters may be determinable, the original script is no longer possible to read. The gradual erasure shares some qualities with the lipogram, the main difference being that in the lipogram, the omission of a character is intentional. In Feurstein's (2024) analysis of the phenomenon among the German Pietists, they explain that the absence of a letter could signify a lack, while the repetition of one would indicate the opposite. The omission of a letter opens up the document to additional layers of interpretation, but here we run once again into the question of difference and negation. Is the omission really just that: A difference? To use a suitably blunt example, in the etching of script into a clay tablet, it is not only the etching that is the text. Rather, it is that space, the relief into the object, where the text arises. The inversion, ink on paper, also relies on the relief around the ink in addition to its contrast with the paper. And so in the case of the lipogram, it would seem to suggest that its signifying power comes not from being a site of difference, but negation. Actual difference, as between two unique characters, has a different material effect than omission in that the script cannot arise without absence. With the lipogram, that omission is, of course, intentional, but with the hunger stones, the eroding of a character, while probably expected when created, is not intentionally done to produce hermeneutical complexity. Yet it does, and this is where that erosion becomes a positive feature of the document. As the document changes, its own capacity as a harbinger becomes intensified until, at some point in total illegibility, all that can be read is a sort of non-specific warning, in a way more troubling in its openness towards possible crises.

As it pertains to the extinction of the multispecial, the document finds itself positioned between the necessity for documenting the past while pointing forward as it does so. What was initially viewed as a type of inertia owes primarily to the illusion of its paralysis as a kind of event divorced from time. By emphasising decay in its construction, it is able to meet the MUC, with its own aftershock being absorbed by time, on the same plane. As with the case of the hunger stones, the emergence of the document over time and under certain conditions creates a sort of semiotic reversal as the signifying act takes on new meanings with each showing and as it changes from the battering of the elements. No longer is the permanence of the record the ideal lens with which to view it. Rather, its decay affords it opportunities to position itself towards substantive reconsiderations of the necessary mediums that can present different pathways for signification. As one hunger stone reads, 'When this goes under, life will become more colourful again' (Henley, 2022).

Event Horizons

The question of how to document the extinction of the multispecial is not one with a singular answer. In this chapter, I have endeavoured to show the importance of negation in understanding this particular extinction event and how it must necessarily shape our approach to documenting it. This challenge is compounded by the persistence of the colonial, dominating forces driving the Anthropocene, for which such a documentary programme is at best irrelevant. But to ignore or minimise the political opposition will likely only give additional rein to those dominating forces while the multispecial is increasingly ground under the wheels of this planetary project.

As I have noted elsewhere (Radio, 2020), we may very well be the nonhumans on the other side of the Anthropocene. In this sense, the task of coalition building takes on a different light as the emergence of MUCs appears around us. Just as the concept can never fully identify with its object, so too is there a site of tension between the MUC and the document. Remaining in that tension may allow for a different kind of space to emerge between the human and the multispecial by which alternate significations may arise and with it alternate futures.

Notes

1 'Multispecial' refers to the entangled presence, relationships, and knowledge systems shared across multiple species. In the context of ecological crisis and extinction, the multispecial denotes not only the biological diversity being lost but also the epistemic, cultural, and relational dimensions that arise through interspecies coexistence. The 'nonhuman', a closely related term, is broader in scope and so multispecial is used in this examination to foreground relations with other organisms.
2 Zizek (2015) explains a similar relation with the undead, not the dead, as the true negation of the living.
3 See Hjorland's (2007) for an important critique of this argument.
4 This is an interesting yet uneasy connection, for Heidegger's separation of Being from beings, or that 'there is no Being without entities' (Adorno, 2015), intentionally avoids a confrontation with the material that is necessary for relating to the multispecial. But more troubling is his elevation of Being as a power manifesting in historical events to which beings must submit (his own endorsement of National Socialism comes into focus here) (Adorno, 2019). At a time of ecological annihilation, submission to the programme of the Anthropocene is at odds with the documentary praxis I am advocating here.

References

Abramowicz, M. A., & Kluźniak, W. (2001). A precise determination of black hole spin in GRO J1655-40. *Astronomy & Astrophysics*, *374*(3), L19–L20.

Adorno, T. (2015). *Negative Dialectics*. Bloomsbury.

Adorno, T. (2019). *Ontology and Dialectics: 1960/61*. Polity.

Adorno, T. (2020). *Minima Moralia: Reflections from Damaged Life*. Verso Books.

Bates, M. (2022). A proto-paradigm for information science research. A proto-paradigm for information science. In *Proceedings of CoLIS, the 11th. International Conference on Conceptions of Library and InformationScience*. Vol. 27. Oslo, Norway.

Bates, M. (2026). Researching information behaviour among multiple species. In N. Solhjoo (Ed.), *Multispecies Information Science* (pp. 22–38). Routledge.

Bates, M. J. (1999). The invisible substrate of information science. *Journal of the American Society for Information Science*, *50*(12), 1043–1050.

Bates, M. J. (2006). Fundamental forms of information. *Journal of the American Society for Information Science and Technology*, *57*(8), 1033–1045.

Benjamin, W. (2003). Theses on the Philosophy of history. In Howard Eiland and Michael W. Jennings (Eds.), *Selected Writings/ Walter Benjamin, Vol. 4, 1938–1940*, 265–276. Harvard University Press.

Bernstein, J. H. (2009). Nonknowledge: The bibliographical organization of ignorance, stupidity, error, and unreason: Part one. *KO Knowledge Organization*, *36*(1), 17–29.

Briet, S. (2006). *What Is Documentation?: English Translation of the Classic French Text*. Scarecrow Press.

Buckland, M. K. (1991). Information as thing. *Journal of the American Society for Information Science*, *42*(5), 351–360.

Carpenter, J. R., & Kihara, T. (2021). *This Is Not a Good Sign*. https://not-a-good-sign.com/

Cowie, R. H., Bouchet, P., & Fontaine, B. (2022). The sixth mass extinction: Fact, fiction or speculation? *Biological Reviews*, *97*(2), 640–663.

Croissant, J. L. (2014). Agnotology: Ignorance and absence or towards a sociology of things that aren't there. *Social Epistemology*, *28*(1), 4–25.

Day, R. (2026). The Animal's importance for neo-documentation: A 'copernican revolution' in understanding information. In N. Solhjoo (Ed.), *Multispecies Information Science* (pp. 53–60). Routledge.

ESA, NASA, & Mirabel, F. (2002). Fast-Flying Black Hole Yields Clues to Supernova Origin, 18 November 2002. https://hubblesite.org/contents/news-releases/2002/news-2002-30

Feuerstein, H. (2024). Absence as presence: Semiotic hermeneutics of the absent letters in the writings of German pietists. *The Journal of Religion*, *104*(4), 452–478.

Frohmann, B. (2013). The documentality of Mme Briet's antelope. In J. Packer, S. B. Crofts Wiley (Eds.), *Communication Matters* (pp. 173–182). Routledge.

Giovannoni, E., & Quattrone, P. (2018). The materiality of absence: Organizing and the case of the incomplete cathedral. *Organization Studies*, *39*(7), 849–871.

Hartog, F. (2015). *Regimes of Historicity: Presentism and Experiences of Time*. New York: Columbia University Press.

Hayles, N. K., (1999). *How We Became Posthuman: Virtual Bodies in Cybernetics, Literature, and Informatics*, University of Chicago Press.

Henley, J. (2022, 19 August). Hunger stones, wrecks and bones: Europe's drought brings past to surface. *The Guardian*. Retrieved 23 April 2025. https://www.theguardian.com/world/2022/aug/19/hunger-stones-wrecks-and-bones-europe-drought-brings-past-to-surface

Hjørland, B. (2007). Information: Objective or subjective/situational? *Journal of the American Society for Information Science and Technology*, *58*(10), 1448–1456.

Horowitz, A. (2010). *Inside of a Dog: What Dogs See, Smell, and Know*. Simon and Schuster.
Jameson, F. (2003). The end of temporality. *Critical Inquiry*, *29*(4), 695–718.
Jameson, F. (2020). *Valences of the Dialectic*. Verso Books.
Jarvis, S. (1998). *Adorno: A Critical Introduction*. Polity Press.
Malm, A. (2016). *Fossil Capital: The Rise of Steam Power and the Roots of Global Warming*. Verso Books.
Malm, A. (2018). *The Progress of this Storm: Nature and Society in a Warming World*. Verso Books.
Morton, T. (2013). *Hyperobjects: Philosophy and Ecology after the End of the World*. University of Minnesota Press.
Radio, E. (2020). Documents for the Nonhuman. *Journal of Critical Library and Information Studies*, *3*(1), 1–18. https://doi.org/10.24242/jclis.v3i1.108
Sartre, J. P. (1943). *Being and Nothingness*. Washington Square Press.
Sartre, J. P. (1960). *Critique of Dialectical Reason, Volume I*. Verso.
Solhjoo, N. (2024). The red thread of information across more-than-human love. *Journal of Librarianship and Information Science*. https://doi.org/10.1177/09610006241255418
Žižek, S. (1999). *The Ticklish Subject: The Absent Centre of Political Ontology*. Verso.
Zizek, S. (2015). *Absolute Recoil: Towards a New Foundation of Dialectical Materialism*. Verso.

INDEX

For Product Safety Concerns and Information please contact our EU representative GPSR@taylorandfrancis.com
Taylor & Francis Verlag GmbH, Kaufingerstraße 24, 80331 München, Germany

www.ingramcontent.com/pod-product-compliance
Ingram Content Group UK Ltd.
Pitfield, Milton Keynes, MK11 3LW, UK
UKHW020148200926
475702UK00016B/2604
* 9 7 8 1 0 3 2 9 5 1 5 5 3 *